ARDUINO FOR BEGINNERS

ESSENTIAL SKILLS EVERY MAKER NEEDS

John Baichtal

800 East 96th Street,
Indianapolis, Indiana 46240 USA

Arduino for Beginners: Essential Skills Every Maker Needs

ISBN-13: 978-0-7897-4883-6
ISBN-10: 0-7897-4883-5

Library of Congress Control Number: 2013946136

Printed in the United States of America

First Printing: November 2013

Trademarks

All terms mentioned in this book that are known to be trademarks or service marks have been appropriately capitalized. Que Publishing cannot attest to the accuracy of this information. Use of a term in this book should not be regarded as affecting the validity of any trademark or service mark.

Arduino is a registered trademark of Arduino, www.arduino.cc/.

Warning and Disclaimer

Every effort has been made to make this book as complete and as accurate as possible, but no warranty or fitness is implied. The information provided is on an "as is" basis. The author and the publisher shall have neither liability nor responsibility to any person or entity with respect to any loss or damages arising from the information contained in this book.

Bulk Sales

Que Publishing offers excellent discounts on this book when ordered in quantity for bulk purchases or special sales. For more information, please contact

U.S. Corporate and Government Sales
1-800-382-3419
corpsales@pearsontechgroup.com

For sales outside of the U.S., please contact

International Sales
international@pearsoned.com

Editor-in-Chief
Greg Wiegand

Executive Editor
Rick Kughen

Development Editor
Rick Kughen

Managing Editor
Sandra Schroeder

Senior Project Editor
Tonya Simpson

Copy Editor
Paula Lowell

Indexer
Lisa Stumpf

Proofreader
Sarah Kearns

Technical Editor
Pete Prodoehl

Publishing Coordinator
Kristen Watterson

Book Designer
Mark Shirar

Compositor
Mary Sudul

Contents at a Glance

Table of Contents

About the Author

John Baichtal got his start writing blog posts for *Wired*'s legendary GeekDad blog as well as the DIYer's bible *MAKE Magazine*. From there, he branched out into authoring books about toys, tools, robots, and hobby electronics. He is the co-author of *The Cult of LEGO* (No Starch) and author of *Hack This: 24 Incredible Hackerspace Projects from the DIY Movement* as well as *Basic Robot Building with LEGO Mindstorm's NXT 2.0* (both from Que). Most recently he wrote *Make: LEGO and Arduino Projects* for MAKE, collaborating with Adam Wolf and Matthew Beckler. He lives in Minneapolis, MN, with his wife and three children.

Dedication

For Harold Baichtal
1939–2013

Acknowledgments

I want to thank my loving wife, Elise, for her patience and support; all my hacker friends, for answering my endless questions; my mother, Barbara, for working on the glossary; and my children, Eileen Arden, Rosemary, and Jack, for their curiosity and interest.

We Want to Hear from You!

As the reader of this book, *you* are our most important critic and commentator. We value your opinion and want to know what we're doing right, what we could do better, what areas you'd like to see us publish in, and any other words of wisdom you're willing to pass our way.

We welcome your comments. You can email or write to let us know what you did or didn't like about this book—as well as what we can do to make our books better.

Please note that we cannot help you with technical problems related to the topic of this book.

When you write, please be sure to include this book's title and author as well as your name and email address. We will carefully review your comments and share them with the author and editors who worked on the book.

Email: feedback@quepublishing.com

Mail: Que Publishing
ATTN: Reader Feedback
800 East 96th Street
Indianapolis, IN 46240 USA

Reader Services

Visit our website and register this book at quepublishing.com/register for convenient access to any updates, downloads, or errata that might be available for this book.

Introduction

When you go to a store and buy an electronic gizmo, does it ever occur to you that you could make one yourself? Or even that it would be FUN to make one yourself?

The learning curve can be intimidating. You have to study electronics, learn what all the components do, and how to control them with a microcontroller. To put the components together, you'll have to learn how to solder. To program the microcontroller, you'll have to learn how to code. To make a cool container that holds the electronics, you'll have to master certain workshop skills.

Sound intimidating?

One bit of technology that makes these dreams not only achievable but enjoyable is the Arduino, a small microcontroller board designed to be easy to learn and a breeze to program. It lets you operate motors and take input from sensors, allowing you to build the project you want to!

The goal of this book is to help you create those projects—not just the gizmo, but the enclosure as well. You'll learn workshop skills, familiarize yourself with a ton of tools, build stuff. All of these projects use the easy-to-learn Arduino UNO microcontroller.

This book has been a huge learning experience for me, and I hope it is for you as well. You can build a lot of cool things with an Arduino, and the projects in this book are just the beginning. Good luck and have fun!

What's in This Book

This book is designed to take an absolute beginner and bring him or her up to speed on a large number of topics related to electronics, tools, and programming.

- Chapter 1 is called "Arduino Cram Session" because it drops a bunch of information on Arduinos and electronics—just what you need to start creating!
- Chapter 2, "Breadboarding," walks you through actually creating an electronics project—a laser trip beam!—using a handy piece of equipment called a solderless breadboard.
- Chapter 3, "How to Solder," teaches you how to use a soldering iron to connect electronic components. The chapter's project involves adding an LED light strip to a coffee table.
- Chapter 4, "Setting Up Wireless Connections," introduces you to three different ways that you can control a project with wireless signals. When you're finished learning about that, you can tackle the chapter's project, creating a wireless doorbell.
- Chapter 5, "Programming Arduino," shows you the basics of controlling your Arduino with programs you upload to the board. I'll take you line by line through an Arduino program so you can learn how it works.

- Chapter 6, "Sensing the World," describes a variety of sensors and explains the difference between digital and analog sensors. Chapter 6's project is a mood lamp that changes its colors depending on the environment around it.
- Chapter 7, "Controlling Liquid," shows readers three ways to pump liquid, and then puts one of these techniques to the test by showing how to build a plant-watering robot.
- Chapter 8, "Tool Bin," is a crash course on tools, everything from the ultimate toolbox to what to stock a wood or metal shop.
- Chapter 9, "Ultrasonic Detection," talks about using pulses of inaudible sound to map out obstructions and measure distances. The chapter's project, a cat toy, waggles a pompom intriguingly above your cat's nose when the sensor detects her.
- Chapter 10, "Making Noise," shows you how to make delightful electronic music (also known as noise!) generated by your Arduino. The project shows you how to build a hand-held noisemaker of your very own.
- Chapter 11, "Measuring Time," explains three ways in which the Arduino can keep track of time. Then I show you how to build an "indoor wind chime" that strikes on the hour.
- Chapter 12, "Safely Working with High Voltage," shows you three ways to deal with wall current safely. You'll build a sweet lava lamp controller that starts and stops the lamp on a schedule, plus you can trigger it with a remote control.
- Chapter 13, "Controlling Motors," explains motor control options for the Arduino. Then you'll build a bubble-blowing robot with your newfound skills!

Who Can Use This Book

This book is intended for persons new to making Arduinos. It assumes very little knowledge on the part of the reader; the only mental attributes needed are a sense of curiosity and a desire to tackle new challenges.

How to Use This Book

I hope this book is easy enough to read that you don't need instructions. That said, a few elements bear explaining.

Tip

Tips are helpful bits of advice that will save you time and/or headaches.

TIP

This is a Tip that provides helpful advice that I have learned along the way.

Note

Notes are tidbits of useful information that are helpful, but not mission critical.

NOTE

This is a Note that provides information that's useful, even if it is somewhat ancillary.

Caution

Cautions point out pitfalls and dangers. Don't skip these unless you like breaking things and spending time in the ER.

CAUTION

This is a Caution. You shouldn't skip these! The safety of your hardware, tools, and possibly your flesh depend on it.

Parts Lists

For each project in this book, I provide a shopping list of parts, such as the following, that you'll need to complete it.

PARTS LIST

- Arduino
- Servo (I used a HiTec HS-322HD servo, Jameco P/N 33322.)
- Servo horns (A number of horns come with the HiTec; these should be fine.)
- Chronodot RTC Module
- 1/4" dowel (You'll need about 8" to a foot.)
- Wind chime (I used a Gregorian Chimes Soprano wind chime, SKU 28375-00651.)
- 5mm plywood for the enclosure

- 1" pine board for the support blocks
- Eye bolt and nut (The Home Depot P/N 217445)
- #8 × 1/2" wood screws
- #6 × 2" wood screws
- #4 × 1/2" wood screws
- 24 1/4" × 1 1/2" bolts with locking washers and nuts
- 12 1/4" × 1" bolts with locking washers and nuts
- Drill press and a variety of drill bits
- Chop saw
- Table saw

Code

When a project requires code—or a sketch—I list it exactly as you should type it. However, unless you just like typing, you don't need to re-key the code found in this book. I've placed the code online so that you can easily download it, and then copy and paste it. Chapter 5 will get you up to speed on programming your Arduino.

Go to https://github.com/n1/Arduino-For-Beginners to download this code and other files associated with this book.

Here is a sample code listing:

```
int valve = 13; // renames Pin 13 "valve"

int offhours = 0; // how many hours before the water dispenses?
int offmins = 1; // how many minutes before the water dispenses?
int spray = 10; // number of seconds the water sprays

void setup() {
  pinMode(valve, OUTPUT);  // designates the valve pin as "output"
  Serial.begin(115200);
}

void loop() {
  int wait = (offmins * 60000) + (offhours * 3600000); // computes milliseconds

  digitalWrite(valve, HIGH);
  delay(spray * 1000); // water stays on this number of milliseconds
  Serial.println(offmins * 60000); // I used this when debugging
  digitalWrite(valve, LOW);
  delay(offmins * 60000); // water stays off this number of milliseconds
}
```

1

Arduino Cram Session

What do you have to know to successfully create the projects in this book? It turns out, quite a lot! The good news is that I dedicate Chapter 1 to getting you ready to hack. This chapter consists of overviews of basic electronics, tips about workshop safety, as well as coverage of the Arduino Uno itself. Let's get started!

Arduino Uno: A Rookie-Friendly Microcontroller

What if you could shrink a computer down so it would fit on a single circuit board smaller than a playing card? Wouldn't it be awesome if you could add some sensors to detect the environment nearby, buttons to trigger commands, and motors to move stuff? Actually, this isn't a "what if" phenomenon. The device I just described is a microcontroller-based prototyping platform called Arduino.

Want an example of what it can do? Figure 1.1 shows Hexy the Hexapod, a cool robot built by ArcBotics (arcbotics.com) that uses the Arduino platform as its basis. It packs 20 servos and moves using inverse kinematics, a robotics concept that simplifies movement with the use of pre-built routines such as "walk forward." A device as advanced as Hexy the Hexapod certainly is a far cry from making a light blink!

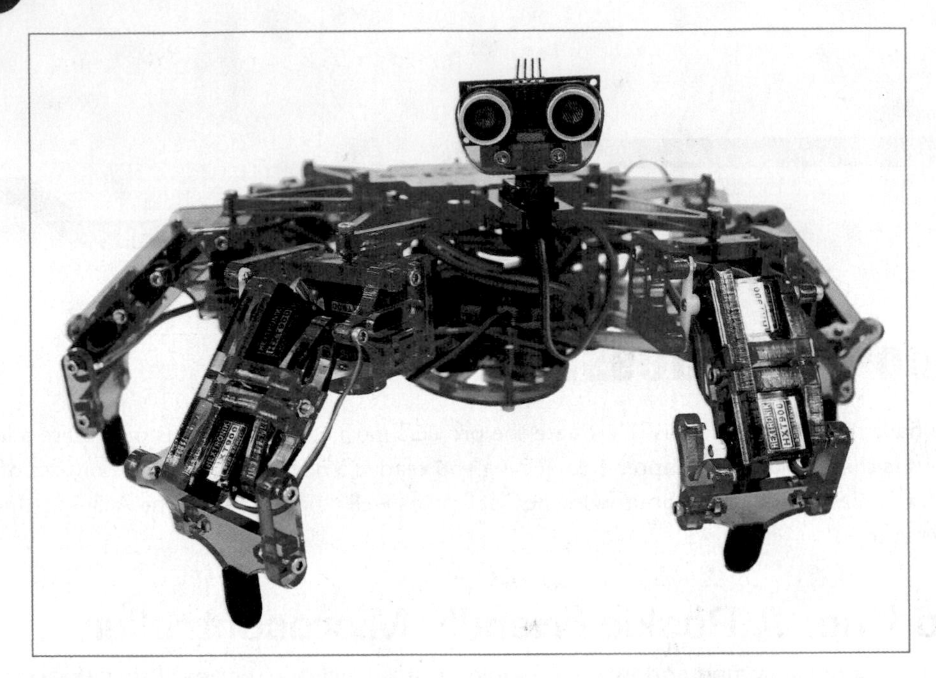

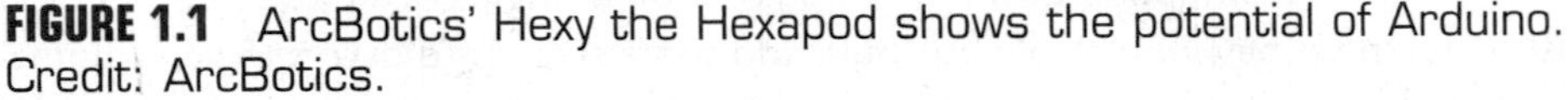

FIGURE 1.1 ArcBotics' Hexy the Hexapod shows the potential of Arduino. Credit: ArcBotics.

Although it's cool, Arduino is by no means the first hobbyist microcontroller. A bunch of others predated Arduino, but none of them have found success the way Arduino has. What originally made the phenomenon a hit is the fact that no competing board is as easy to use. In fact, it was designed specifically for artists, college students, and other casual tinkerers who didn't care to learn a professional environment and just wanted to hack.

Now that the Arduino platform has begun to mature, we're seeing a snowball effect where so many projects, books, websites, and hardware are being developed for Arduino that justifying the use of anything else is hard. This diversity of resources has made the platform itself more visible, which brings in more projects and participants and makes the whole experience more valuable for everyone.

TIP

We Use the Uno

This book makes exclusive use of the Arduino Uno in the projects described. Although many variants and versions of the Arduino exist, the Uno is the default board of the Arduino line, so I focus on it in this book. Some of the other Arduinos are bigger and have more capabilities; others are smaller and stripped down. Not only does the Uno fit nicely in the middle, it is considered by most to be the default model. Chapter 8, "Tool Bin," describes some of the other models.

What exactly do you get when you buy an Arduino? Let's do a quick overview of the board and its features (see Figure 1.2).

FIGURE 1.2 The Arduino is the size of a credit card but is packed with cool possibilities!

1. ATmega328 Microcontroller
2. Pinouts
3. Reset Button
4. Power Jack
5. USB Jack
6. Power Indicator
7. Data Indicators

The Arduino Uno consists of a printed circuit board (PCB) with a microcontroller chip and various other components attached to it. Refer to Figure 1.2 to identify each of the primary components:

- **ATmega328 Microcontroller**—The brains of the board, the ATmega328 features 32KB flash memory, 2KB SRAM, and a clock speed of 16 MHz. This might not sound robust, but Arduino programs are quite small.

- **Pinouts**—You'll attach wires to these little ports. For example, you could plug in a button to one and a motor into another. Some of them do different things than others, and we explore these differences later on in the book.
- **Reset button**—When all else fails, press this button. It restarts the Arduino and automatically relaunches whatever program is loaded onto it.
- **Power jack**—This power jack can accommodate a nine-volt AC adapter ("wall wart") with a 2.1mm, center-positive plug. You also can connect an ordinary nine-volt battery, as long as it has been equipped with the same plug. We explore the various ways of powering your Arduino projects in Chapter 8.
- **USB jack**—This jack accepts power and data from a standard A-B USB cable, the sort that is often used for printers and other computer peripherals. Not only is this cable used to program the Arduino, it also powers the board, so you can prototype a project without wasting batteries.
- **Power indicator**—This LED lights up when the board has power.
- **Data indicators**—These LEDs flash when data is being uploaded to the board.

TIP

Downloading the Arduino Software

You won't learn how to program the Arduino in this chapter, but you can get started by downloading the software. It's compatible with Windows, Macintosh, and Linux and doesn't cost a penny. Go to http://arduino.cc/en/Main/Software and follow the directions you see there. If you need more help, I walk you through the download process in Chapter 5, "Programming Arduino."

Other Arduino Products

As mentioned, an entire ecosystem of Arduino variants exists, as well as add-on circuit boards called shields. The Arduino variants include more powerful Arduinos for major projects, small ones for small projects, and shields—add-on boards—that do everything from playing music to connecting to the Internet to determining GPS coordinates.

Often, a project you're contemplating has already been tackled by someone else, who made it into a shield. If you're looking to add a certain capability to your project, first consider looking for an existing shield—it could save you a great deal of work! Even better, some shields can be stacked on top of each other, allowing you to build progressively more complicated assemblies.

The Relay Shield (see Figure 1.3) is an example of a shield shown stacked on top of an Arduino. Created by DIY gurus Evil Mad Science LLC, the shield uses a relay to control a high-voltage circuit. For example, the shield could be used to safely trigger a lamp that uses wall current. You can buy the Relay Shield kit at http://evilmadscience.com/productsmenu/tinykitlist/544.

FIGURE 1.3 Evil Mad Science's Relay Shield plugs into the Arduino's pinouts and allows it to control high-voltage circuits.

Interested in learning more about shields? Be sure to check out Chapter 8, where we examine a number of shields and other add-on boards.

Electronics

An Arduino is cool, but you'll need some electronic components to make it do anything interesting! The project shown in Figure 1.4 uses LEGO motors and an aquarium pump controlled by an Arduino to make and dispense chocolate milk. In Chapter 7, "Controlling Liquid," I show you how to make a similar pump. In the meantime, the following sections provide a brief overview of some of the more commonplace components that you'll encounter.

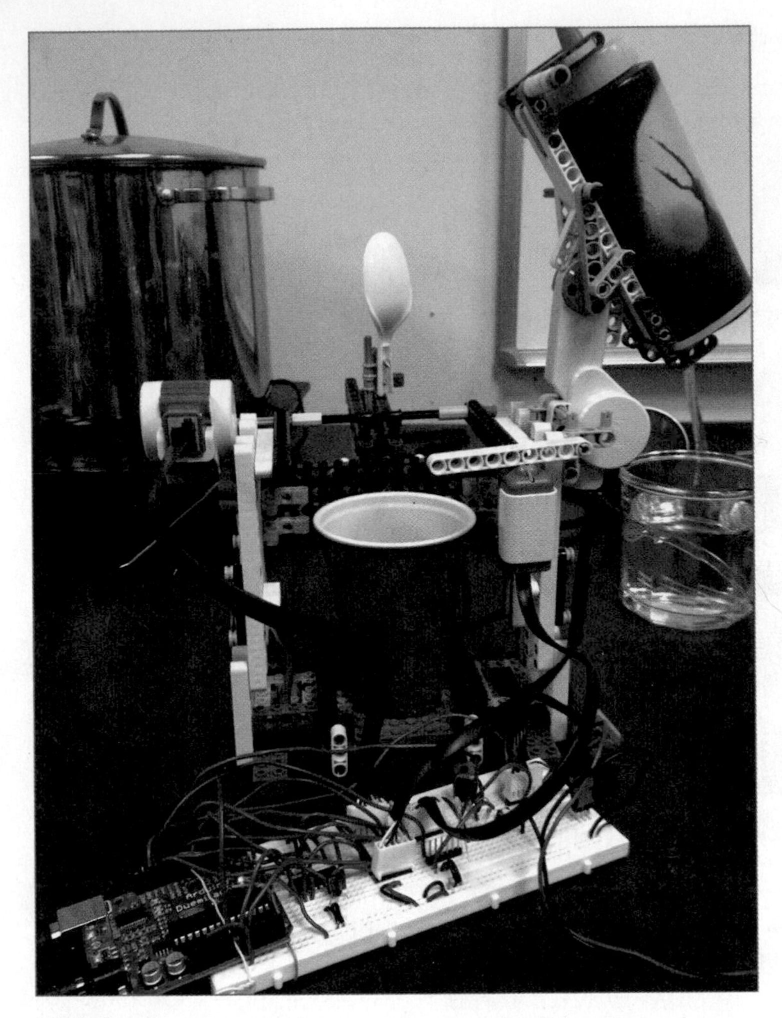

FIGURE 1.4 You need more than an Arduino to make a cool project.

NOTE

This Is Just an Overview

Many other varieties of components are available for you to learn, some of which I cover later in the book. Others you might have to learn about on your own. My goal in the following sections is to get you started with the basics.

Light-Emitting Diodes

LEDs (see Figure 1.5) are the lights of the Arduino world. They come in different colors and intensities, and some include additional features, such as blinking or the capability to change color based on the software inputs. LEDs that can change color are called RGB (red-green-blue) LEDs, and you use them later in this book.

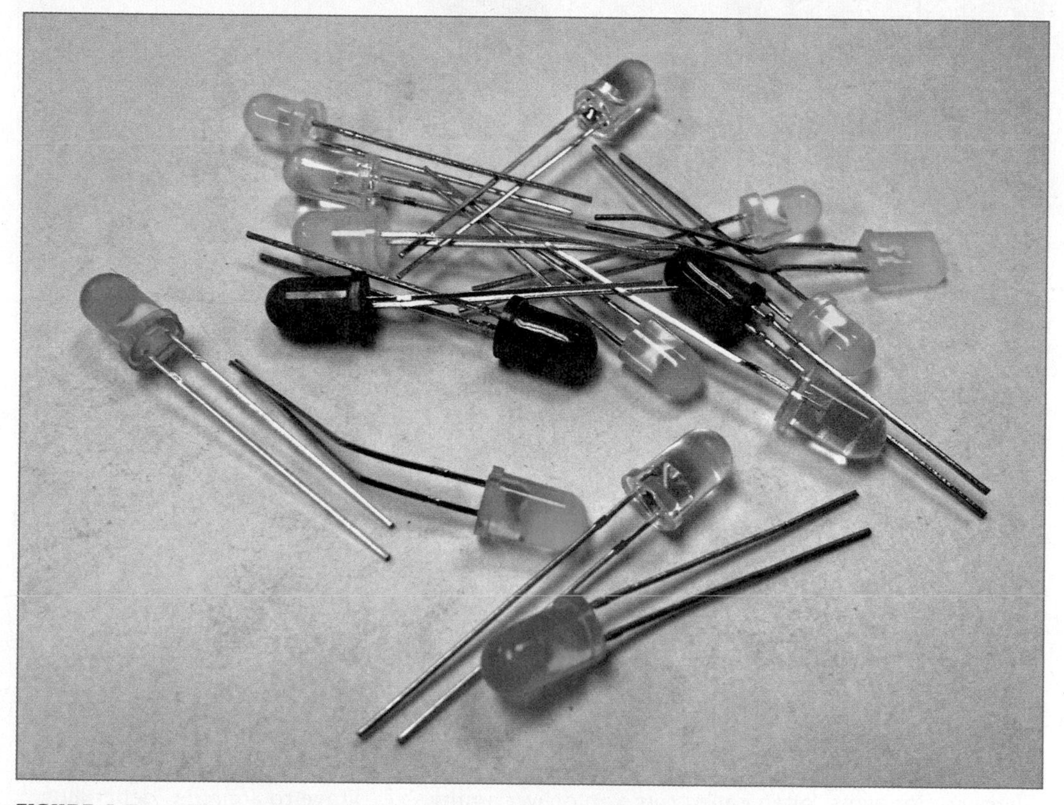

FIGURE 1.5 LEDs are the light bulbs of the electronics world.

Buttons and Switches

Arduino responds to human inputs with the help of buttons and switches, as shown in Figure 1.6. An amazing variety of these components exist, which is good! This enables you to find exactly the right configuration for whatever project you're working on. You can do all sorts of fun things with switches, such as using two subprograms on your Arduino and toggling between the two when the switch is thrown.

FIGURE 1.6 Buttons and switches tell the Arduino what you want it to do next.

Potentiometers

These components, often called *pots*, can deliver a range of voltage to a circuit, depending on how far the knob is turned. For example, you could make an LED shine brighter if you turn the knob one way or dimmer if you turn it the other way. Pots can control data as well. For example, you could program in different behaviors depending on how the pot is turned. Many different sizes and shapes of pot exist, as you can see in Figure 1.7.

FIGURE 1.7 Potentiometers enable you to control a circuit with the turn of your wrist.

Resistors

Electricity is the friend of electronic components, right? Well, yes, but too much juice can damage them. That's where resistors come in. These small components stop all but a fraction of the electricity from passing through to the component. Resistors are rated in ohms. The resistors shown in Figure 1.8 are most commonly used in hobbyist projects. They are marked with color bands so you can identify how many ohms each resistor has. You can find a guide to the color codes in Chapter 8.

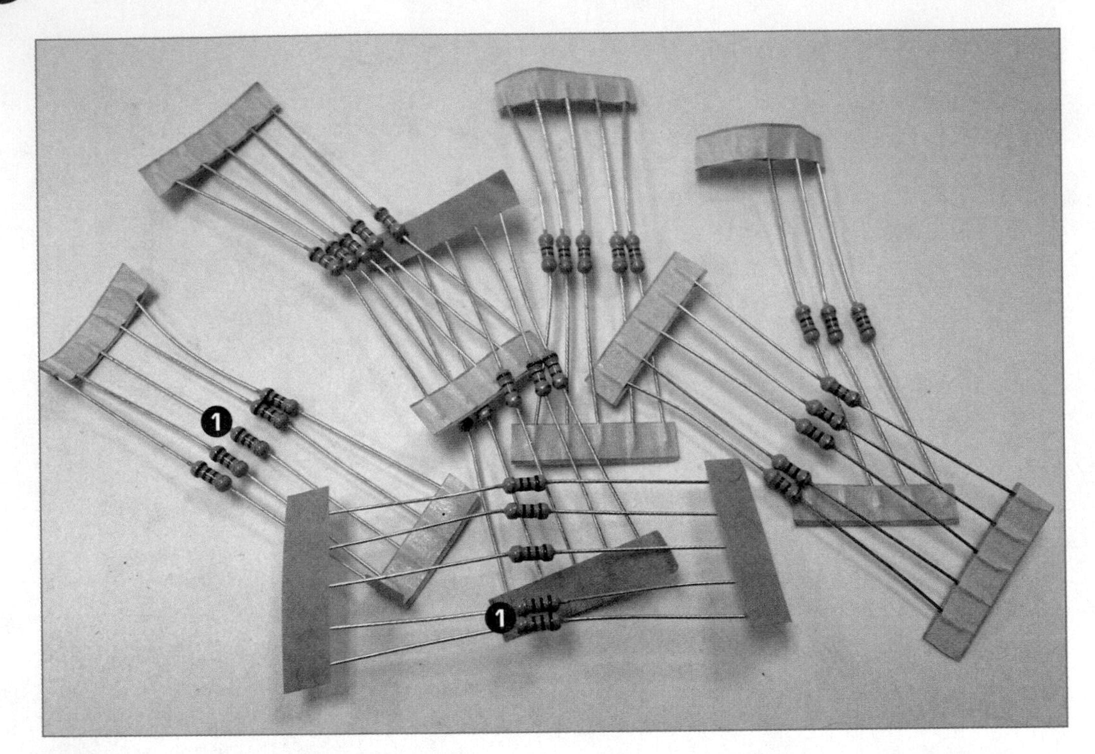

FIGURE 1.8 Resistors limit the flow of electricity, protecting your components from too much juice.

1 Colored bands identify the ohms.

Capacitors

Capacitors (often just referred to as *caps*) store and discharge small amounts of electricity, enabling them to be used as timing devices because, when paired with resistors, a cap discharges at a predictable rate. Because of this predictability, capacitors are also used to "clean up" an electronic signal, such as the frequency response of an audio circuit. Figure 1.9 shows a variety of capacitors.

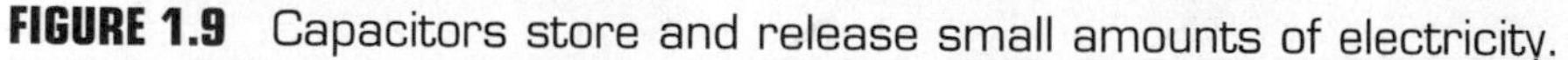

FIGURE 1.9 Capacitors store and release small amounts of electricity.

Motors

As shown in Figure 1.10, the following are the three main types of motor that you'll learn about in this book:

- **Steppers**—A stepper motor rotates in "steps" rather than just rotating willy-nilly. This enables you to control its movement precisely, and therefore it is used for computer-controlled milling and other tasks that require control down to the millimeter.
- **Servos**—Servos are motors that have "encoders" built in that send position information back to the microcontroller. Servos are often used for robots where control of the motors' shafts is important but not critical.
- **DC motors**—DC motors have no feedback or other means of control beyond the application of electricity. When a charge exists, the motor turns. When the charge stops, the motor does as well. DC motors are used in projects where the shaft's position really doesn't matter at all, such as in a remote-controlled helicopter.

FIGURE 1.10 Steppers, servos, and DC motors comprise the main types of motors used in hobby electronics.

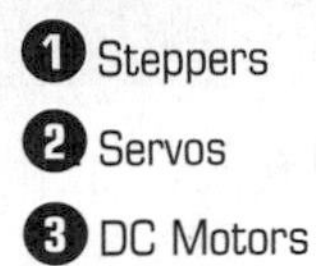

Solenoids

Whereas motors turn, a solenoid (see Figure 1.11) uses an electromagnet to move a shaft back and forth. One application for this is a valve; when the right voltage passes through the solenoid's coils, the valve opens. When the voltage stops, the valve closes.

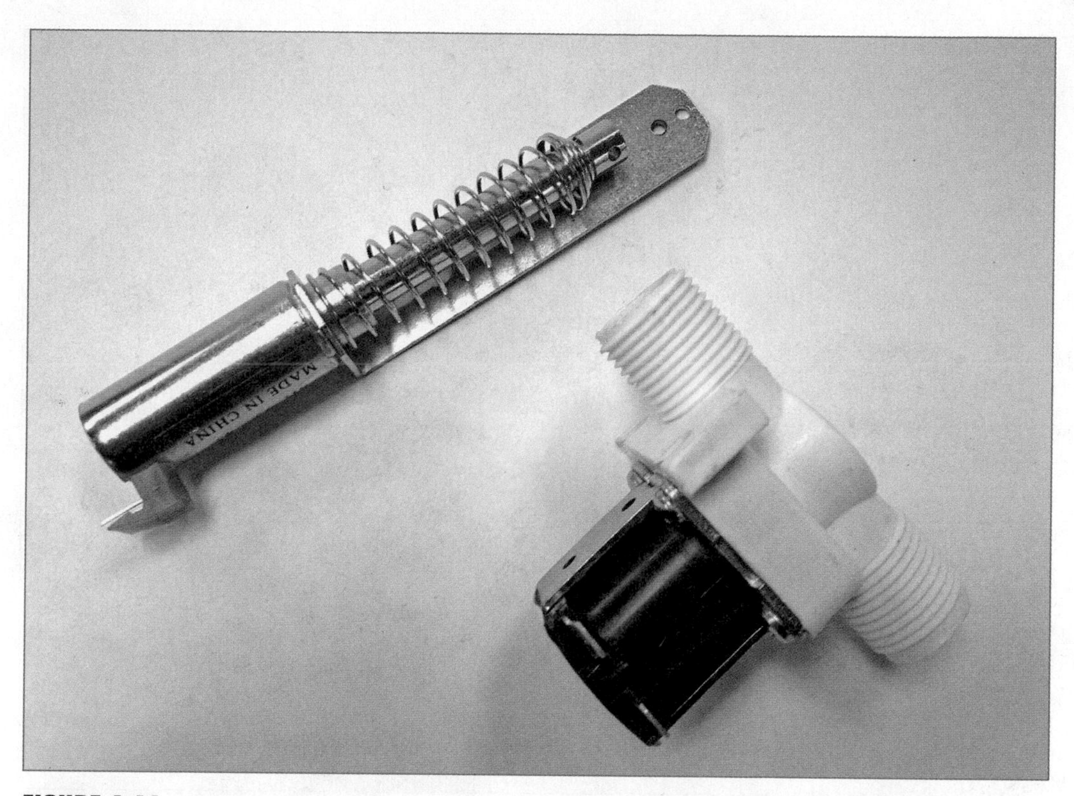

FIGURE 1.11 Solenoids are like motors but move the shaft back and forth instead of rotating it.

Piezo Buzzers

The primary noisemakers used in electronic kits are piezos, shown in Figure 1.12. You apply voltage, and a noise comes out. Pretty simple!

FIGURE 1.12 Want to create a buzz with your Arduino? Plug in one of these piezos.

Seven-Segment Displays

Say you want a display in your project that shows letters and numbers. The classic solution is a seven-segment display that consists of a number of LEDs (usually seven) that can be selectively lit up to show you a letter or number. Many different styles are available, as shown in Figure 1.13, but they mostly work the same way.

FIGURE 1.13 A variety of displays exist; each consists of a series of LED segments that can be triggered individually to create letters and numbers.

Relays

Relays (see Figure 1.14) are like electronic switches: When your program sends a triggering current to the relay, it activates another circuit. For example, if you wanted to control a lamp that uses wall current, you could use a relay paired with an Arduino to control the lamp's current without needing to handle AC current yourself! Figure 1.3 earlier in this chapter shows an Arduino shield that controls a relay.

FIGURE 1.14 Relays serve as Arduino-controlled switches, triggering circuits on command.

Integrated Circuits

Also known as ICs, integrated circuits (see Figure 1.15) are just what they sound like: entire circuits packed into individual chips, simplifying your electronics projects—assuming you can find the right IC! Examples of ICs include microcontrollers, such as the ATmega328 used in Arduinos, timer chips, amplifiers, and so on.

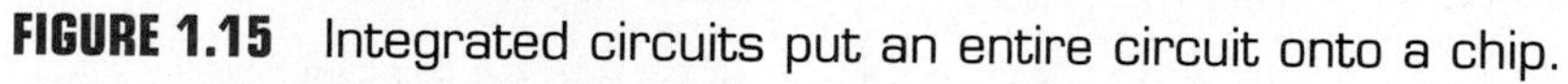

FIGURE 1.15 Integrated circuits put an entire circuit onto a chip.

Temperature Sensors

The temperature sensor (see Figure 1.16) takes in information about the temperature in the area and returns a value to the Arduino. This sensor is a great addition to such projects as weather stations or for triggering a cooling fan, for example.

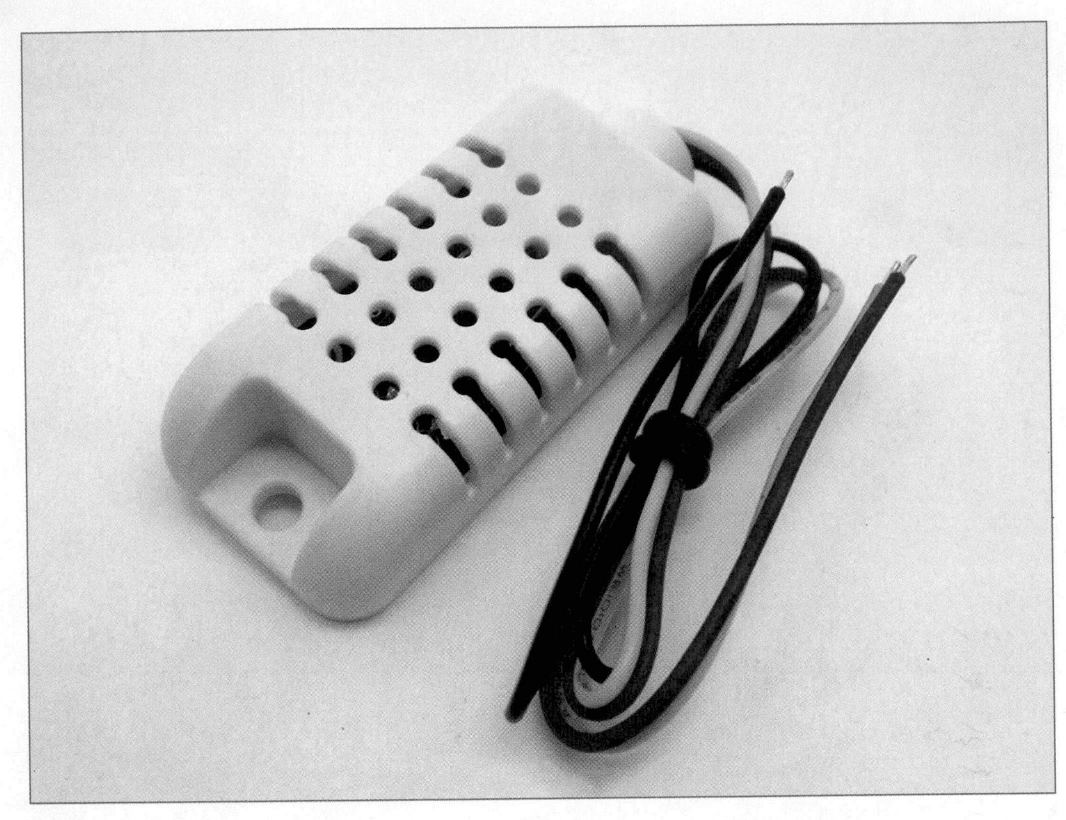

FIGURE 1.16 Temperature sensors tell the Arduino about the environment around it.

Flex Sensor

Great for wearable electronics, the flex sensor (see Figure 1.17) tells the Arduino when it is bent by changing the amount of electricity allowed to pass through it. Wouldn't it be great to control a robot hand with a flex sensor-equipped glove?

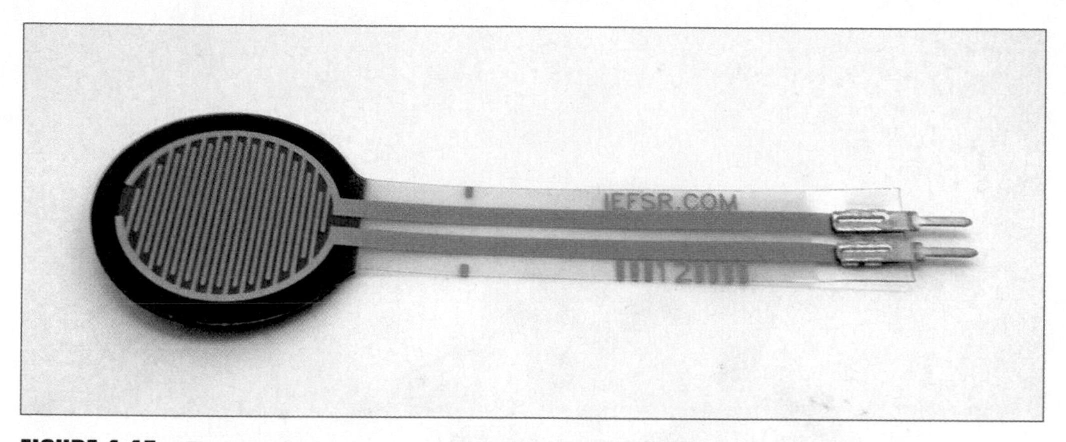

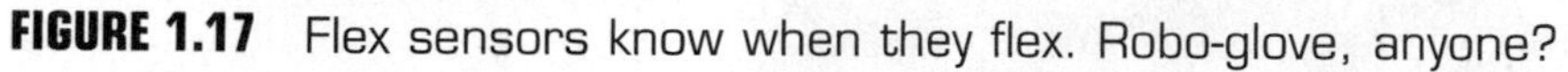

FIGURE 1.17 Flex sensors know when they flex. Robo-glove, anyone?

Light Sensor

Light sensors (see Figure 1.18) are often used in electronics projects. In fact, you'll use them a few times in this book! Basically, the sensor tells the Arduino how light or dark it is, triggering different events depending on the light level.

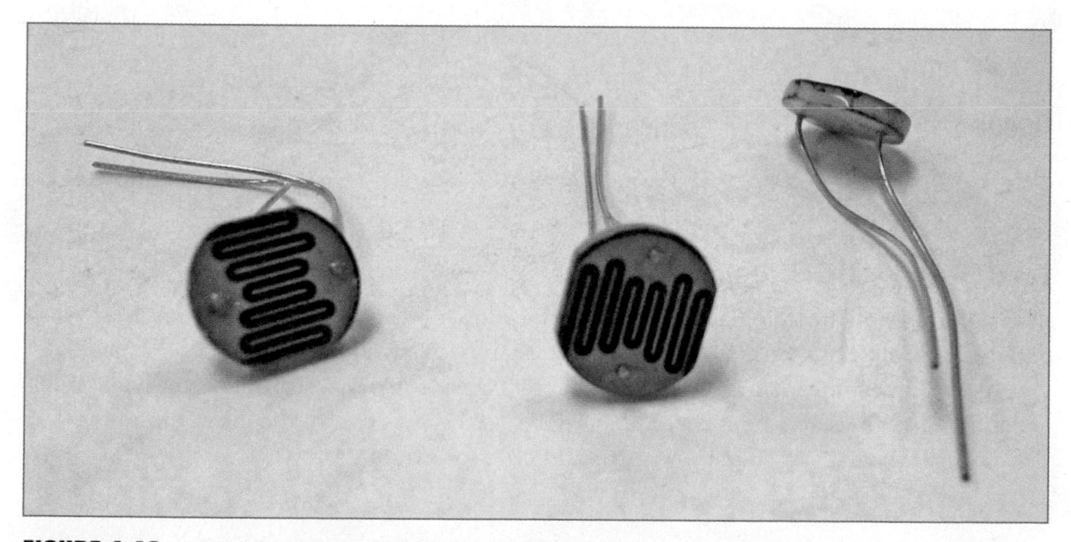

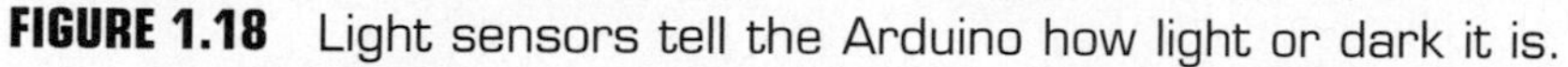

FIGURE 1.18 Light sensors tell the Arduino how light or dark it is.

Ultrasonic Sensor

Ultrasonic sensors (see Figure 1.19) detect movement nearby by beaming out inaudible—to humans!—pulses of noise, while listening for the noise to bounce back. This is kind of how a bat's echolocation works.

FIGURE 1.19 The ultrasonic sensor sees by bouncing ultrasonic pulses off of nearby objects.

Safety Rules

In this book, you'll be doing a lot of work in the shop, and this means using tools that could potentially hurt you. This section provides an overview of some basic shop safety rules that apply in most situations. Later chapters cover some more specific situations you'll encounter and the safety rules that apply. Figure 1.20 shows two of the most important pieces of safety gear you should own—ear and eye protection!

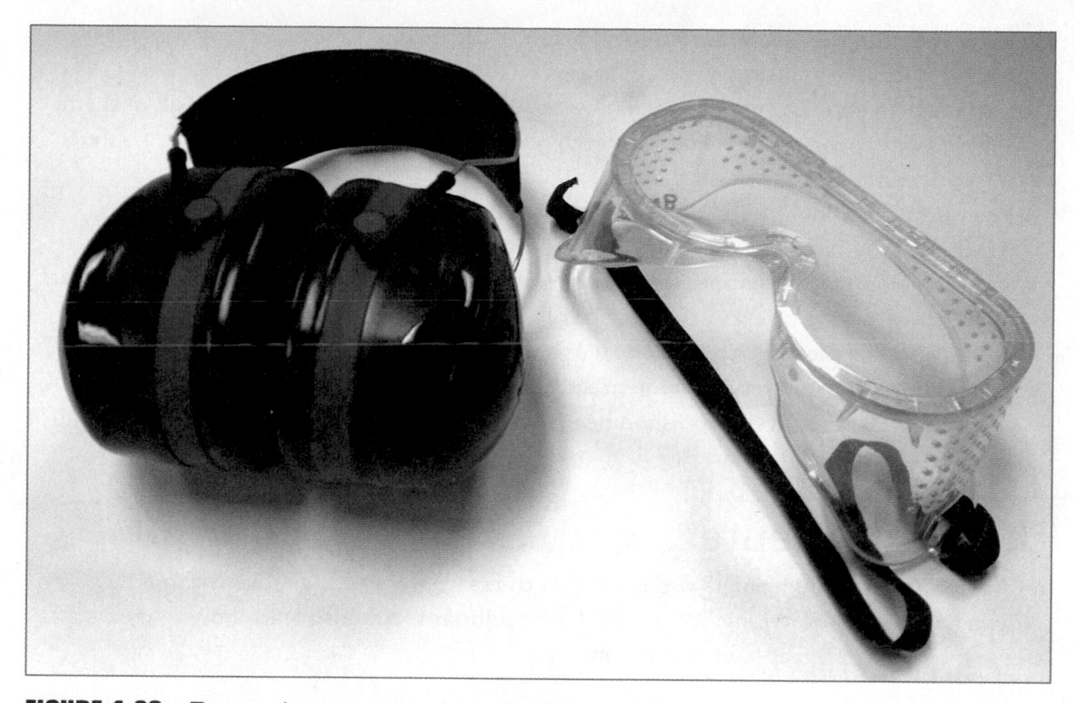

FIGURE 1.20 Ear and eye protection should not be neglected.

Follow these rules:

- **Use protection**—Goggles, hearing protection, dust masks, and protective clothing are often necessary, depending on what tool you're using. If you're using irritants, you'll need skin protection. If you have long hair and are using a power tool, pull your hair back so it doesn't get caught. Always use goggles if there's a chance that something will fly into your eyes; regular eyeglasses are not good enough.
- **Be aware and alert**—Stay away from drugs and alcohol, especially when using power tools. Furthermore, make sure to get plenty of sleep—many a maker have gotten hurt while pulling all-nighters.
- **Cleanliness is important**—If you're working by yourself, the temptation might be to let your workshop get messy. Don't do it! You're more likely to have an accident in a messy shop than in a clean one.
- **Be aware of your surroundings**—Know who is in the workshop with you and where they are in proximity to you and the tool you're using. For instance, if you're using a power saw and a friend drops a wrench with a loud clang, an injury could result.
- **Know your tool**—You should be respectful of your tools but not scared of them. If you're using a new tool, learn about it first. Either ask an experienced maker to "check you out" or, if you don't know someone like that, you can often find YouTube videos demonstrating how the tool is used. Similarly, use the tool for its intended purpose. Many people have been injured using a screwdriver as a pry-bar, for instance.

- **Keep your tools in good condition**—If a saw blade is dull, for instance, you might have to "force it" when cutting, which increases your chance of injury.
- **Know where your fingers are**—You have ten of 'em—ideally—and you need all of them. When using power saws, welders, or even regular hammers, make sure you're aware of the danger and keep your digits safe.
- **Keep a first aid kit**—In addition to the usual stuff like alcohol swabs, adhesive bandages, and tweezers, be sure to stock gauze pads and tape in your kit because maker injuries can sometimes be serious. Also, saline eyewash squeeze bottles are great for getting irritants or even sawdust out of your eyes. Chapter 8 provides complete descriptions of the ultimate maker's first aid kit.
- **Don't forget basic safety equipment such as fire extinguishers and smoke detectors**—A sink is great, too. Every workshop needs ready access to a sink to wash off irritants or to rinse a wound.

The Next Chapter

In Chapter 2, "Breadboarding," you learn how to create electronic circuits without soldering, using a prototyping board called a breadboard. You also learn how to create a laser trip beam to protect your home from intruders!

2

Breadboarding

In this chapter, you learn all about *breadboarding*, the quick and easy way to prototype Arduino projects. After you're up to speed on that, you'll tackle your first project: a laser trip beam for your house! You'll also learn how to use a passive infrared sensor in place of the laser, as well as how to design and cut a plywood enclosure for your trip beam.

Assembling Circuits Using Solderless Breadboards

Sure, you could make your project a tangle of wires, but sometimes a little organization can make a project easier to understand. Often, for everyone from newbies to experts, the first step toward building a project is to breadboard it. Look at the project in Figure 2.1. If the creator wants to make a change, it's incredibly easy—the work of seconds. It's perfect for prototyping.

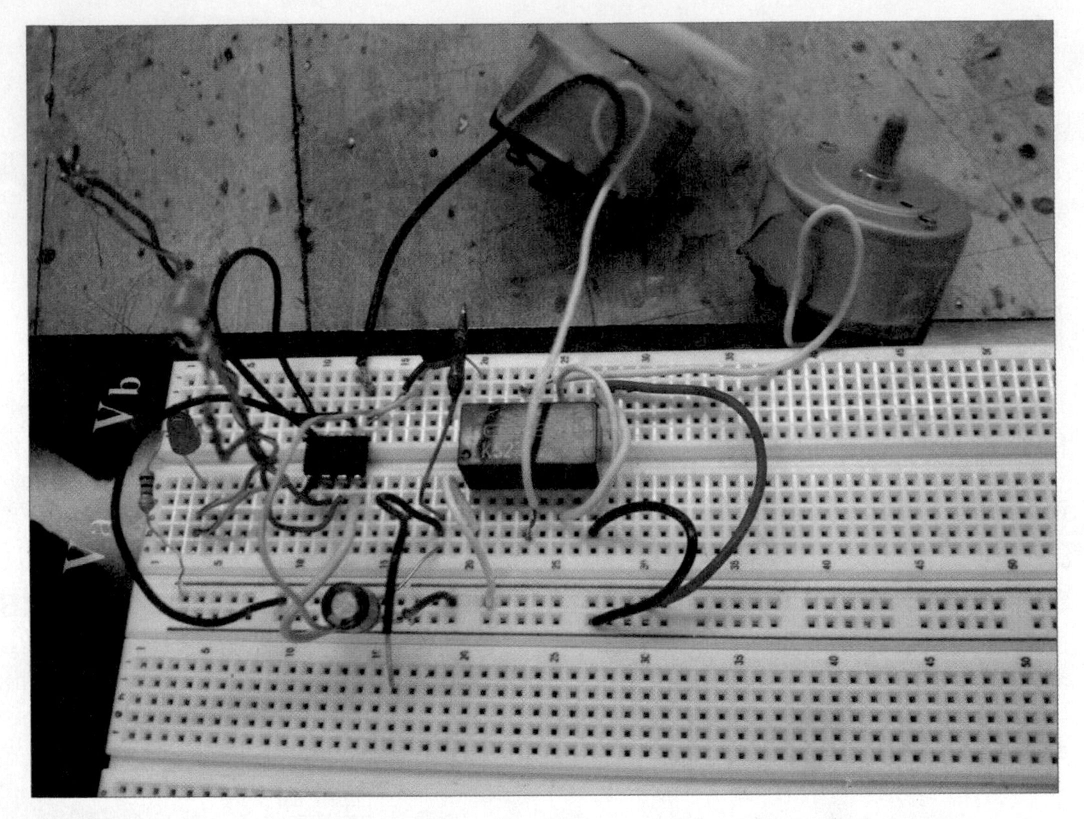

FIGURE 2.1 A breadboarded project is easy to set up and modify because you don't have to solder! Credit: Chris Connors

A solderless breadboard is a plastic board covered in wire holes and featuring concealed conductors. These conductors in essence serve as additional wires for your project. You simply plug in your Arduino, motors, sensors, and so on to the board and use it to manage the connections. Figure 2.2 shows how a breadboard works.

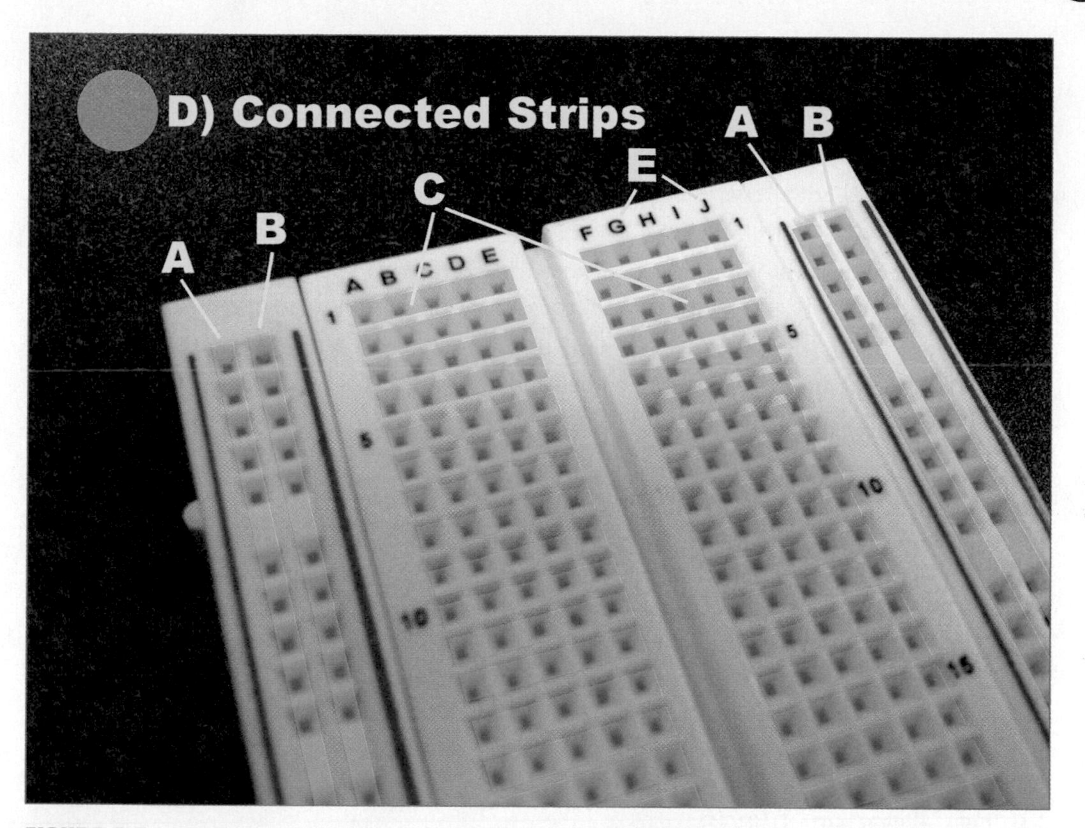

FIGURE 2.2 How does a breadboard work? This photo shows the operation of a typical breadboard.

The following list describes how each of the connections functions:

A. Ground bus strip—Connect to a GND pin on your Arduino. Ground strips are usually marked in blue or black.

B. Power bus strip—Connect a power supply to power the strip. Note, however, that the two strips aren't connected. Power strips are usually marked in red.

C. Terminal strips—The terminal strips are perpendicular to the bus strips. Note that I have marked the terminal strips with light blue shading so that they stand out in Figure 2.2. Your breadboard will not be shaded in this way.

D. Conductors—The blue strips indicate where the concealed connectors are positioned.

E. Hole letters and numbers—These help you describe your project. For example, "Plug the wire into H4" means you would find Row H and then count down to the fourth plug.

Typical breadboards consist of two bus strips on each side, with a power strip, usually marked in red, as well as a ground strip marked in blue or black. Perpendicular to the bus strip are the terminal strips. These are the ones in the middle, and consist of short rows of wire holes linked together by hidden conductors as marked in blue in Figure 2.2.

Conductors are essentially wires, kept hidden so that your project is easier to wire up. Trying to decipher a huge tangle of wire is a lot to ask of a beginner.

To use these conductors, you simply plug in a wire to the row you want, and then plug in the component or wire to which you want to connect to another part of the row. It's easy!

The last thing you need to know about solderless breadboards is that many of them have an adhesive on the back. This feature comes in handy in this chapter because you'll use the adhesive to stick the breadboard to the enclosure.

Understanding Power and Ground

Without getting into how electricity works too much, let's cover two important terms you'll find used a lot in electrical projects: power and ground. Put simply, in circuits, the power wire is where electricity comes from, and the ground wire is its return path. On a breadboard, both power and ground get their own bus strips, allowing you to easily power individual elements of the circuit. You'll plug in your power supply—whether from the Arduino or a secondary supply—to the power bus strip, and connect the ground bus strip to one of the GND pins on your Arduino.

Using Jumper Wires

The wires typically used in breadboarding projects are called *jumpers*. You can create your own simply by using wires clipped to the right length. You'll want to avoid stranded wire because it has a tendency to fray; use solid wire instead.

Alternatively, you can purchase specially created jumper wires. These consist of a slender wire—in a variety of colors and lengths—with a connector pin attached to each end, as shown in Figure 2.3. The pins are designed to fit perfectly into the holes of a solderless breadboard, while being durable enough to be reused many times.

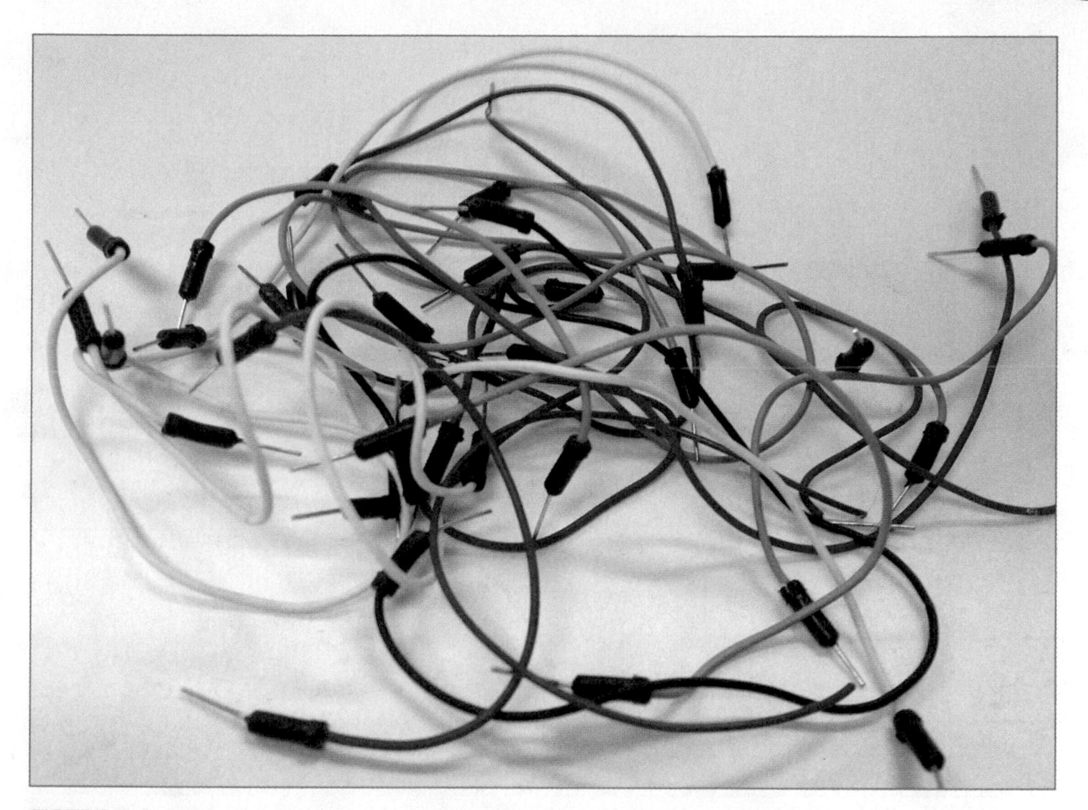

FIGURE 2.3 You can use practically any wires on your breadboard, but wires with pins already attached, as shown here, work the best.

Another type of jumper wire comes in pre-cut lengths, already angled so that you can simply slip them into the breadboard's holes and the wires lay flat against the board, organizing what otherwise would be a confusing tangle of wire. You can see this type of jumper in Figure 2.4.

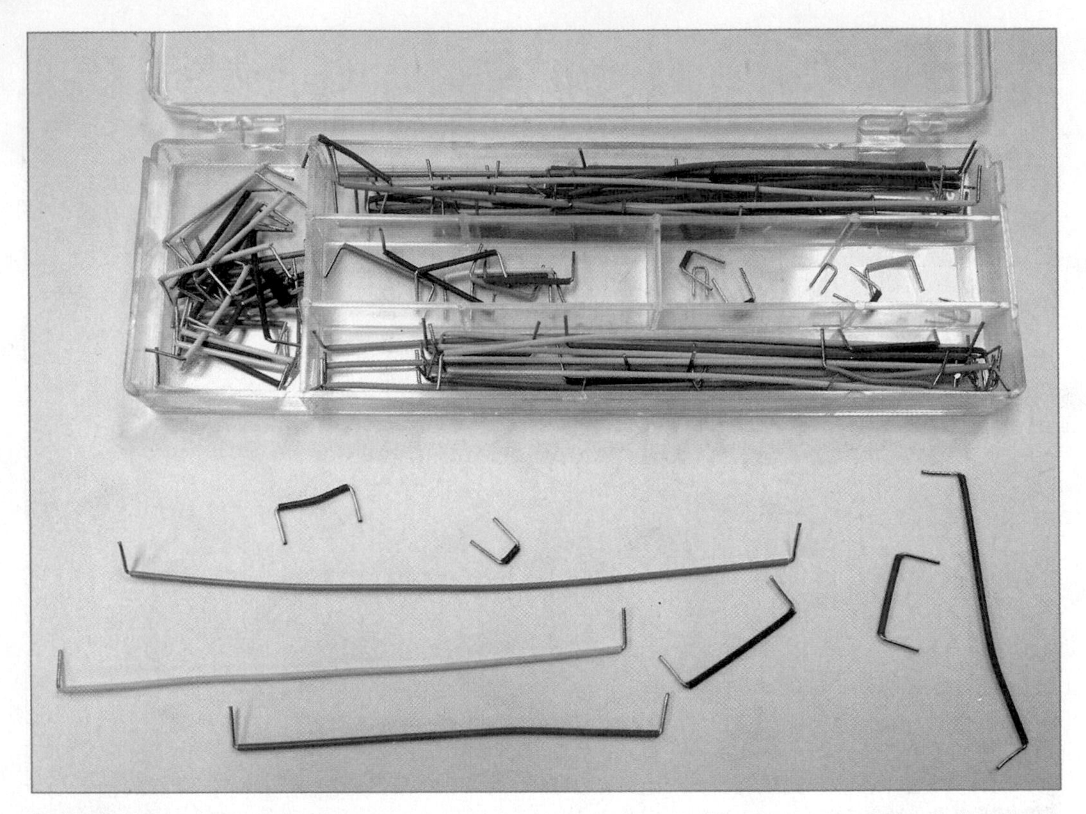

FIGURE 2.4 Breadboard jumpers are great because the pre-cut wires fit flush against the breadboard, keeping your wire neat and orderly.

Before you get started with your main project, let's do a simple breadboarding project to get your confidence up for the real thing.

Project: Breadboard Blink

Here's a simple project you can do in just a couple of minutes; it can help you get up to speed on breadboards. All the project does is light up an LED, as you can see in Figure 2.5. Sound simple? It is.

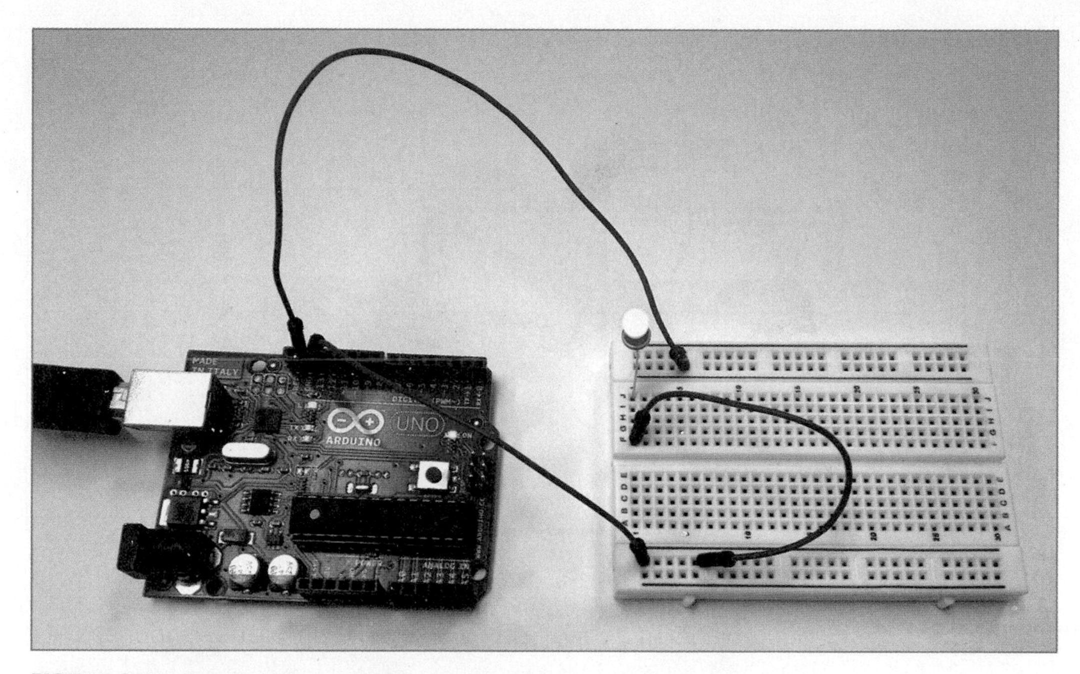

FIGURE 2.5 Do you want to make an LED blink?

PARTS LIST

- **Arduino Uno**—If you haven't bought an Arduino yet, now is the time! In a perfect world, you should use Revision 3 of the Arduino Uno because (at least at the time of this writing) it is the latest and greatest. You can buy it from the Maker Shed (http://www.makershed.com/New_Arduino_Uno_Revision_3_p/mksp11.htm). However, this project works with pretty much any version of Arduino.
- **Breadboard**—I used a half-size breadboard for this project, similar to this one from the Maker Shed: http://www.makershed.com/product_p/mkkn2.htm.
- **USB Cable**—For this project, you use the most common kind, with a flat connector that plugs in to your computer and a square connector that plugs in to the Arduino.
- **Jumper Wires**—You can either use regular ol' wire—anything that can plug in to the breadboard is fine—or buy dedicated wires. The Maker Shed has a decent assortment: http://www.makershed.com/SearchResults.asp?Search=jumper&Submit=Search.
- **An LED**—These are astonishingly cheap. The Maker Shed has an assortment of 100 that costs $8 (http://www.makershed.com/Make_100_LED_Assortment_p/mkee7.htm)!

The following instructions tell you how to set up your breadboard. With this project, and throughout the book, I use an electronic visualization program called Fritzing (fritzing.org) to show you how to wire up your circuits. You can see a sample in Figure 2.6. Chapter 8, "Tool Bin," explores Fritzing in more detail.

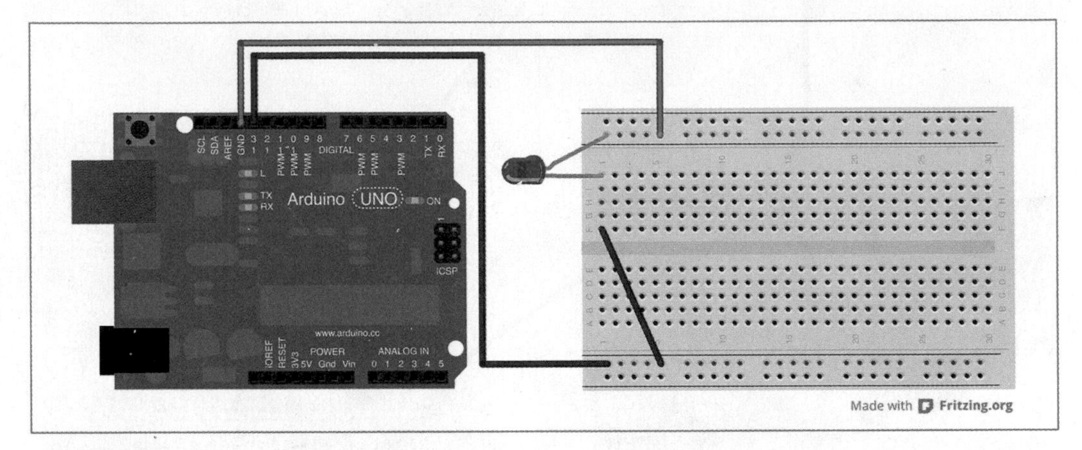

FIGURE 2.6 This Fritzing diagram shows you how to wire up your breadboard.

1. Plug in an LED to the breadboard. The longer lead of the LED goes in to J1 and the shorter lead plugs in to the ground terminal bus right next to it.
2. Plug in a wire to one of the Arduino's GND ports (in Figure 2.6, the wire is plugged in to the GND next to port 13). The other end of the wire can plug in anywhere in the ground terminal bus. I have it in the fifth row.
3. Connect a wire from port 13 of the Arduino to the power (red) bus strip of the breadboard.
4. Connect the bus to the LED's terminal strip as previously shown in Figure 2.3. As you can see, I have the terminal strip connected to F1.
5. If you haven't already, download the Arduino software from this website: http://arduino.cc/en/Main/Software. The web page provides instructions for installing the software. Also, be sure to read Chapter 5, "Programming Arduino," for more instructions. Load the Blink example, found in File, Examples, 01.Basics of the Arduino software.
6. Connect your Arduino to the computer via the USB cable, which will also power your project. Select **File**, **Upload** to send the Blink program to the Arduino.

Voilà! You should have a blinking light on your breadboard.

If it doesn't work, try these troubleshooting steps:

- Test your connections against the preceding instructions.

- Try plugging the LED directly into the Arduino, with the long lead plugging in to port 13 and the short lead plugging in to the GND port next to 13. It should blink if you uploaded the program correctly. Additionally, the tiny LED on the board (labeled "L") will also blink.
- Finally, if you're still having trouble, see Chapter 5, which covers project debugging.

NOTE

More on Programming Later

If these instructions confuse you, never fear! I explore how to program your board in Chapter 5, which guides you through the whole process in more detail.

Project: Laser Trip Beam

The Blink project was just to get your feet wet. Let's do something cool! The main project for this chapter is a Laser Trip Beam (see Figure 2.7). It consists of a small laser that shines into a light sensor connected to an Arduino. If the sensor stops detecting the light, a buzzer sounds the alarm. It's fun, easy, and let's face it—awesome!

FIGURE 2.7 Trip this laser beam and set off a buzzer!

The project consists of two modules:

- The first is the assembly that emits the laser. You'll set up a battery pack, the laser, and the locking switch that arms the laser.
- The other unit consists of an Arduino, a photoresistor, and a buzzer to sound the alarm.

You'll also explore an alternative way to set an alarm using a special sensor called a Passive Infrared (PIR) sensor. It's a motion detector typically used in security systems, and you'll learn how to set up your own.

LASER SAFETY

The laser you use in this project is relatively modest in power—similar to a laser pointer—and won't burn your skin or start a fire. That said, even weak lasers can damage your retinas permanently (see Figure 2.8). Never let a laser shine into your eye, even for a moment.

FIGURE 2.8 They don't make these signs for no reason: A laser can blind you!

PARTS LIST

You'll need the following parts to build your trip beam:

- **Arduino Uno**
- **Power supply for the Arduino**—A "wall wart" rated for 9V with a 2.1mm center-positive plug (such as Adafruit P/N 63, www.adafruit.com/).
- **Battery pack**—See Digi-Key P/N BC22AAW-ND, www.digikey.com/.
- **Sugru**—This is easily moldable putty that cures into rubber; see Chapter 8 for more information. You can buy it at www.sugru.com.
- **Heat-shrink tubing**—See Anytime Tools P/N 201263, www.anytimesale.com.
- **Laser card**—See Sparkfun P/N COM-00594. Figure 2.9 shows the laser card I used for this project.

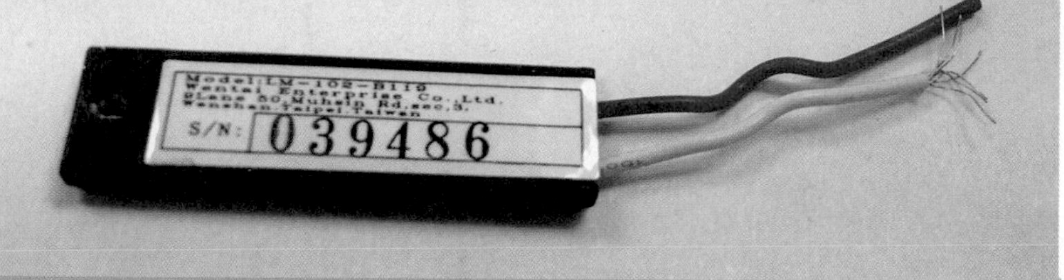

FIGURE 2.9 This laser card is rated for 0.8mW and is powered by 3 volts.

- **11mm photo resistor**—See Digi-Key P/N PDV-P5003-ND, www.digikey.com/.
- **10K resistor**—See Jameco P/N 691104, www.jameco.com; must be bought in sets of 100.
- **A half-size breadboard**—See Adafruit P/N 64, www.adafruit.com/.
- **Two keylock switches**—See Digi-Key P/N EG2625-ND, www.digikey.com/.
- **Buzzer**—See Jameco P/N 1956741, www.jameco.com.
- **Wire**
- **Breadboard jumpers**
- **Standoffs**—3/8 inch (Sparkfun P/N 10461, www.sparkfun.com)
- **Machine screws**—#4-40 × 1"

Let's build it! You begin with the laser module (diagrammed in Figure 2.10) because it's relatively simple. When you're finished with that, you'll move onto the sensor module.

Assembling the Laser Module

To assemble the laser module, follow these steps:

1. Connect the battery pack's red wire to one terminal of the keylock switch, as shown in Figure 2.10. It doesn't matter which terminal. See the sidebar on how to heat-shrink a wire to the lock's terminals, later in this chapter.

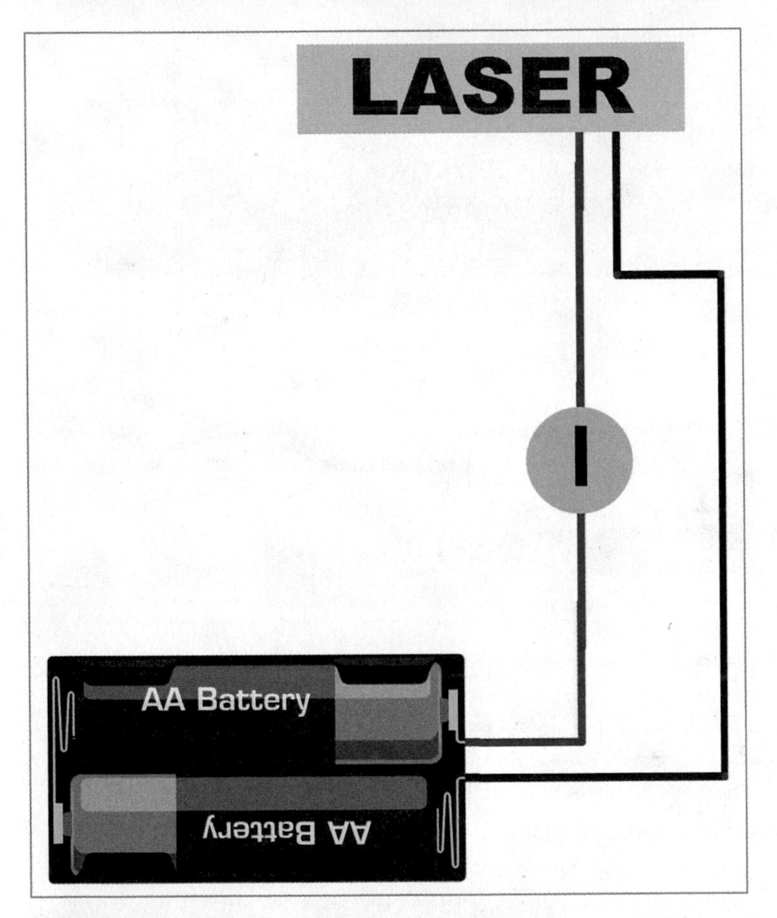

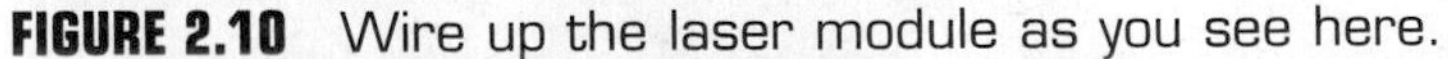
FIGURE 2.10 Wire up the laser module as you see here.

2. Connect the other keylock terminal switch to the laser's red wire.
3. Connect the black wire of the battery pack to the laser's black wire.

Assembling the Sensor Module

Now move on to building the sensor module (see Figure 2.11) as follows:

1. Connect the GND port of the Arduino to the ground bus of the breadboard.

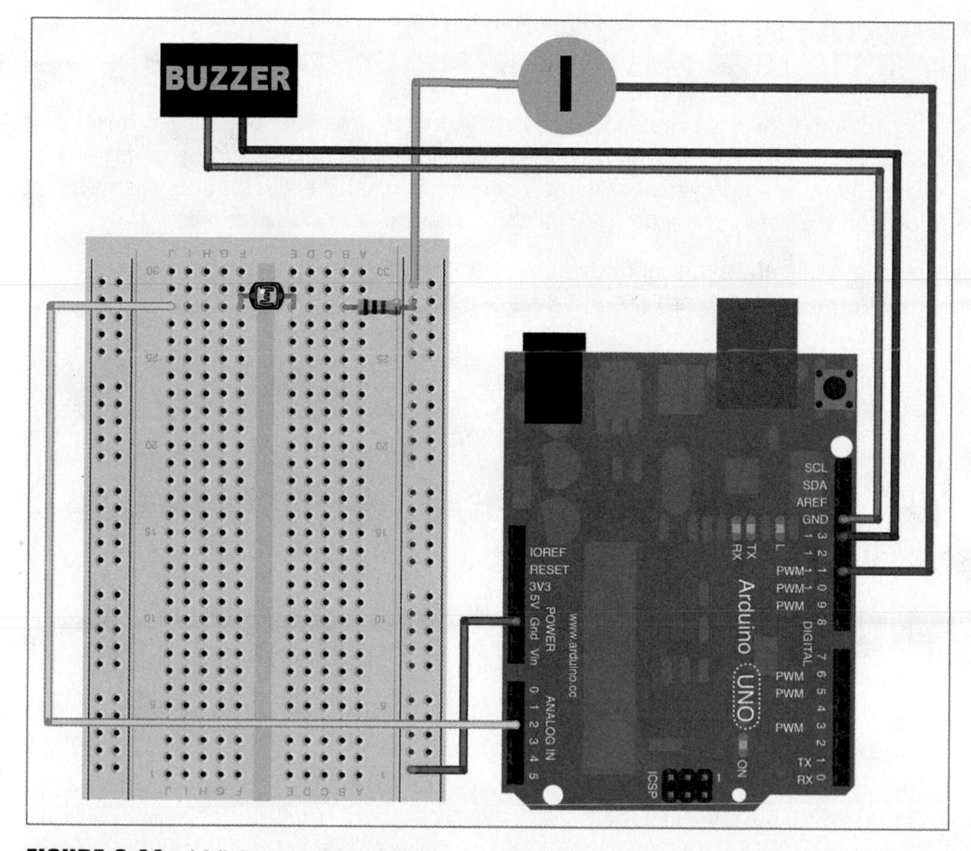

FIGURE 2.11 Wiring up the sensor module is more complicated than the other one, but still not too difficult!

2. Plug in the photoresistor and the resistor to the breadboard, as shown in Figure 2.11. Note that one end of the resistor plugs in to the ground bus.
3. Connect the buzzer to the Arduino. The black wire plugs in to GND, and the red wire plugs in to port 13. (Note that I use different colored wires in Figure 2.11 to help differentiate the wires; you need not use wires of these colors unless you really want to.)
4. Add the keylock switch, shown as a gray circle with a line. One terminal connects to port 11 and the other connects to the ground bus of the breadboard.

HOW TO USE HEAT-SHRINK TUBING

Heat-shrink tubing (HST) is a great product for electronic tinkerers because it helps keep wires connected to their terminals. Basically, HST is a rubbery tube that fits around a wire, and then contracts to form a secure fit when heat is applied. Here's how you can use the tubing to attach wires to the keylock switch terminals:

1. Strip about a half inch of insulation from one end of a wire, and thread about an inch of heat-shrink tubing onto the wire, as shown in Figure 2.12.

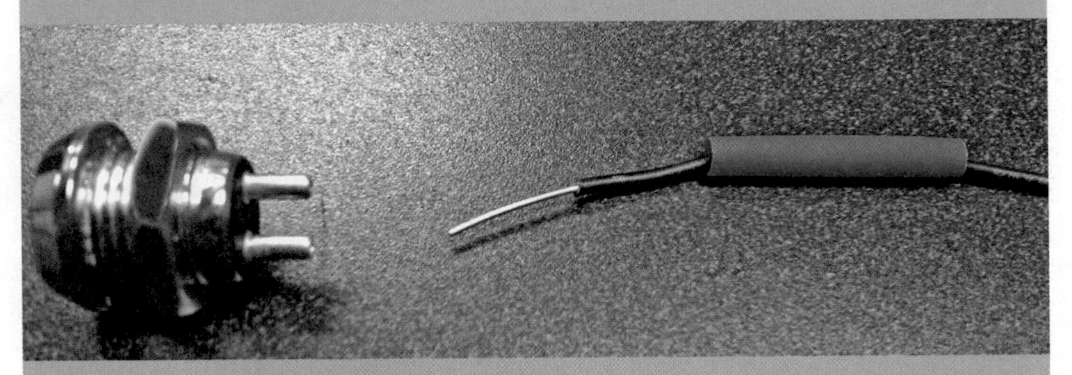

FIGURE 2.12 The keylock switch's terminals are smooth posts. You need heat-shrink tubing to connect the wire.

2. Wrap the exposed end of the wire around one of the terminals of the key lock, keeping it as tight as possible. Wrap it a few times more with the insulated part of the wire (see Figure 2.13).

FIGURE 2.13 Wrap the wire around the terminal several times.

3. Pull up the tubing so it covers the terminal and wrapped wire.
4. Apply a source of heat, such as a crème brulée torch, soldering iron, or heat gun. Be careful not to burn your fingers or ignite any flammable items on your bench. The tubing contracts and securely holds the wires in place. When you're finished, it should look just like Figure 2.14. Now, do the other terminal the same way.

FIGURE 2.14 Secure the wire by contracting the tubing around it.

Chapter 3, "How to Solder," shows you how to do this connection using a soldering iron and solder. This method is even more secure than heat-shrink alone!

Building the Enclosures

The next step in this project is to build the enclosures. I designed the boxes in Adobe Illustrator and output the panels on a laser cutter. At their most basic level, the enclosures are just wooden boxes. The main difference with this design is that it uses quarter-inch teeth that nest with other panels, creating a remarkably solid container for your project after it's finished.

The panels' teeth (shown in Figure 2.15) equal the thickness of the material, and when paired with the precision of a laser cutter, you get a set of panels that connect so perfectly that they barely need glue at all: Friction keeps them together. That said, you still glue all the panels but one, and you secure that final panel with screws.

FIGURE 2.15 Panels from one of the enclosures, ready to be assembled into a box.

To make the enclosures:

1. Download the patterns from https://github.com/n1.
2. Use a laser cutter to cut them out of quarter-inch MDF (medium-density fiberboard). For my settings I used a speed of 10, a power of 100, and a frequency of 1,000 on a 35-watt Epilog.

LASER CUTTER ALTERNATIVES

What, you say? You don't have access to a laser cutter? Chapter 8 includes a tutorial on how you can operate one of these cool gadgets, but in the meantime, you might just want to create your own boxes out of wood pieces you cut yourself, repurpose a cardboard box, or buy a project enclosure from an electronic hobbyist's store.

3. After you have the pieces cut, glue the first side to the base, as shown in Figure 2.16.

FIGURE 2.16 Begin the gluing process by gluing one side to the base.

4. Next, glue the remaining pieces except for the back; you must keep that panel removable to add the electronics. I suggest gluing the pieces by adding a drop to each tooth, as shown in Figure 2.17 (I probably used too much glue). Wipe up the excess glue and leave it to dry. Note that you probably don't need to clamp or secure it; those laser-cut parts fit together very snugly!

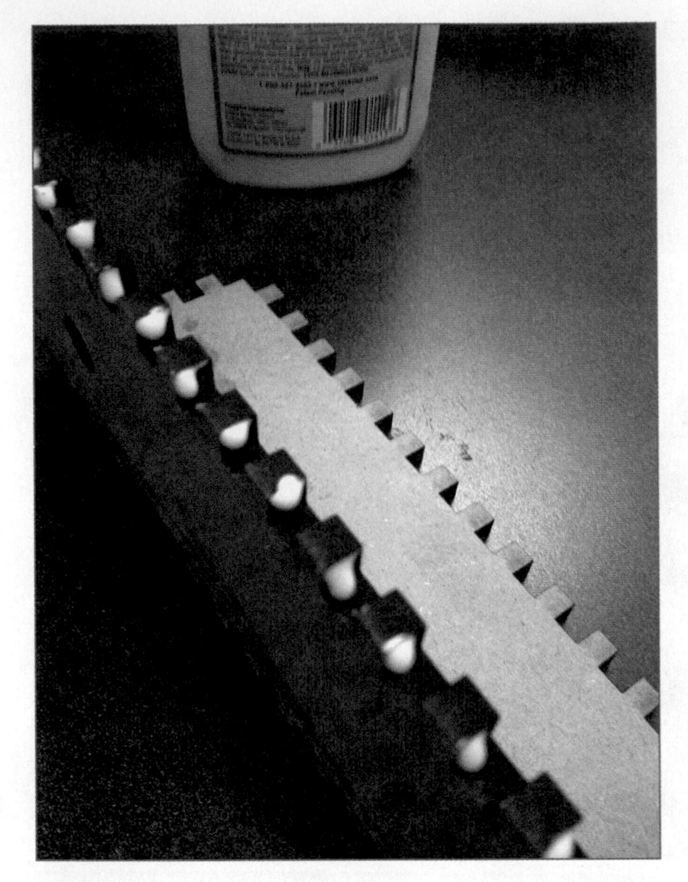

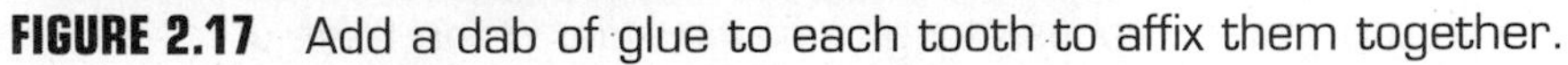

FIGURE 2.17 Add a dab of glue to each tooth to affix them together.

5. So, you have your finished enclosures; they should look like the ones shown in Figure 2.18. You might want to paint them at this point.

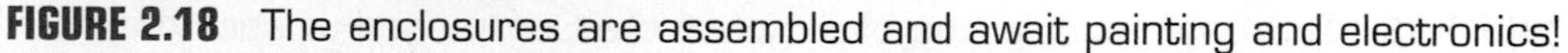

FIGURE 2.18 The enclosures are assembled and await painting and electronics!

Now it's time to add electronics to the laser module enclosure, which is the smaller one of the two. You already assembled the guts in steps 1–4, so it'll be a breeze!

1. You can let the battery pack rattle around at the bottom of the enclosure, or you can simply screw or hot glue the pack to the inside of the box.
2. Thread the keylock switch through the top hole of the enclosure and tighten the nut. If the switch wants to rotate, you might want to hot glue the switch in place.
3. Glue a piece of wood to the inside of the box as shown in Figure 2.19. I chose a piece of wood about 0.75" in length. Use a piece of Sugru putty (described in Chapter 8) to attach the laser to the wood so the beam shines through the hole in the front of the enclosure. Be sure the laser is shining exactly how you want it, then let the Sugru cure overnight; this holds the laser in place.

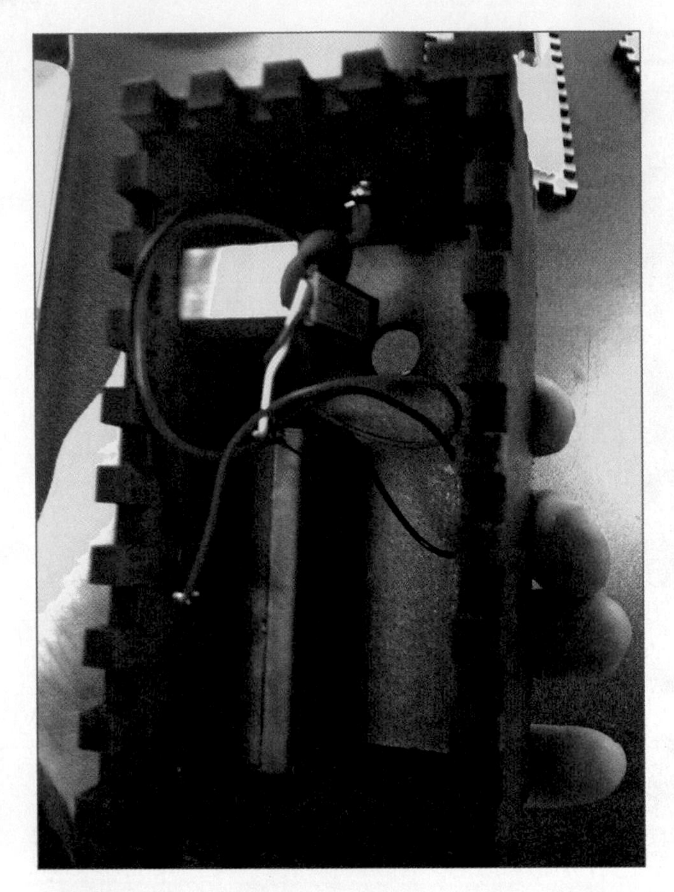

FIGURE 2.19 The laser module; the blue blob is the Sugru used to stick the laser to the block of wood.

4. Attach the back panel to the laser module enclosure, and secure it with some slender wood screws.

Now add the electronics to the sensor module enclosure:

1. Attach the Arduino to the front panel of the enclosure using the #4-40 screws and the standoffs. (See Figure 2.20, which shows the enclosure from the back.)

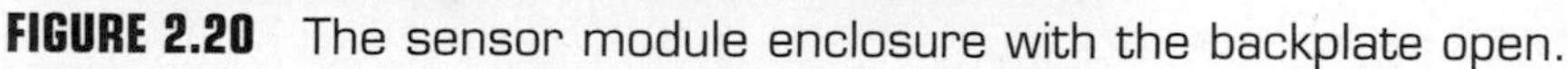

FIGURE 2.20 The sensor module enclosure with the backplate open.

2. Screw on the buzzer, keeping it close enough to the Arduino to connect the two together.
3. Peel off the adhesive backing on the breadboard and stick it to the backplate. Be sure to leave enough room for the power supply.
4. Add the keylock switch to the top hole as you did with the laser module.
5. If you wired all the components as explained earlier in this chapter, you should be set! Be sure you have the photoresistor positioned so that it's visible through the front hole when the module is assembled.

Laser Trip Beam Code

Use the following code to program your Arduino. Note that you'll want the latest version of the Arduino software installed or an error might result. You can find it at arduino.cc.

You can download the trip beam code at https://github.com/n1/Arduino-For-Beginners. Not sure how to upload code to the Arduino? Read the first part of Chapter 5 to learn how!

```
//these tell the Arduino which pins will be used in the program
#define sensorPin A2
#define buzzerPin 13
#define keylockPin 11
int sensorValue = 0;
int threshold = 0; //make this number higher if the alarm trips too readily
void setup() {
  //this part declares whether each pin is an input pin or output pin
pinMode(keylockPin, INPUT_PULLUP);
   pinMode(buzzerPin, OUTPUT);
   pinMode(sensorPin, INPUT);
   Serial.begin(115200);
}
void loop() {
  //this loop arms the alarm based on the status of the keylock switch
  while (digitalRead(keylockPin) == HIGH) {
    senseIntruder();
  }
}
void senseIntruder() {
  //this function compares the light sensor's value against the threshold
  //to see if the beam has been interrupted
  int sensorValue = analogRead(sensorPin);

  Serial.println(sensorValue); //for debugging purposes

  if (sensorValue > threshold) {
    digitalWrite(buzzerPin, HIGH);
      delay(2000); //this sets the duration of the alarm. Higher # = longer buzz
  }
  else {
    digitalWrite(buzzerPin, LOW);
  }
  delay(20);
}
```

Setting Up the Trip Beam

Now that you've completed the two modules, it's time to set them up (Figure 2.21 shows my completed laser beam).

1. Find a door or hallway that you want to secure, and then set up the two modules to shine the beam across the pathway, ensuring that the laser beam hits the light sensor.

FIGURE 2.21 The trip beam in place. Yes, the photo is blurry. You try taking a picture of an invisible beam of light in a dark room!

2. Find an outlet for the sensor module, ideally with the module actually covering the outlet so that it can't be easily unplugged, and plug it in.

   If you can't find a good outlet, another option might be to plug in a 9V battery to the Arduino to power it; Adafruit has a convenient battery pack (P/N 67) with a barrel plug that connects to the Arduino's DC plug.
3. After the enclosures are set up, turn the key on the laser module to activate the beam.
4. Turn the key on the sensor module. The beam is now armed!

You need to make two adjustments based on the ambient lighting in your room:

- If the alarm goes off too readily, you must change the threshold in the code.
- If the alarm doesn't go off enough, try a different resistor on the breadboard; instead of a 10K, try a 5K. This gives the light sensor more range on the lower end.

Alt.Project: Infrared Detector

Obviously, the trip beam is not a serious security measure. Another way, arguably more effective but less cool, is to use an infrared sensor (see Figure 2.22) to detect the intrusion.

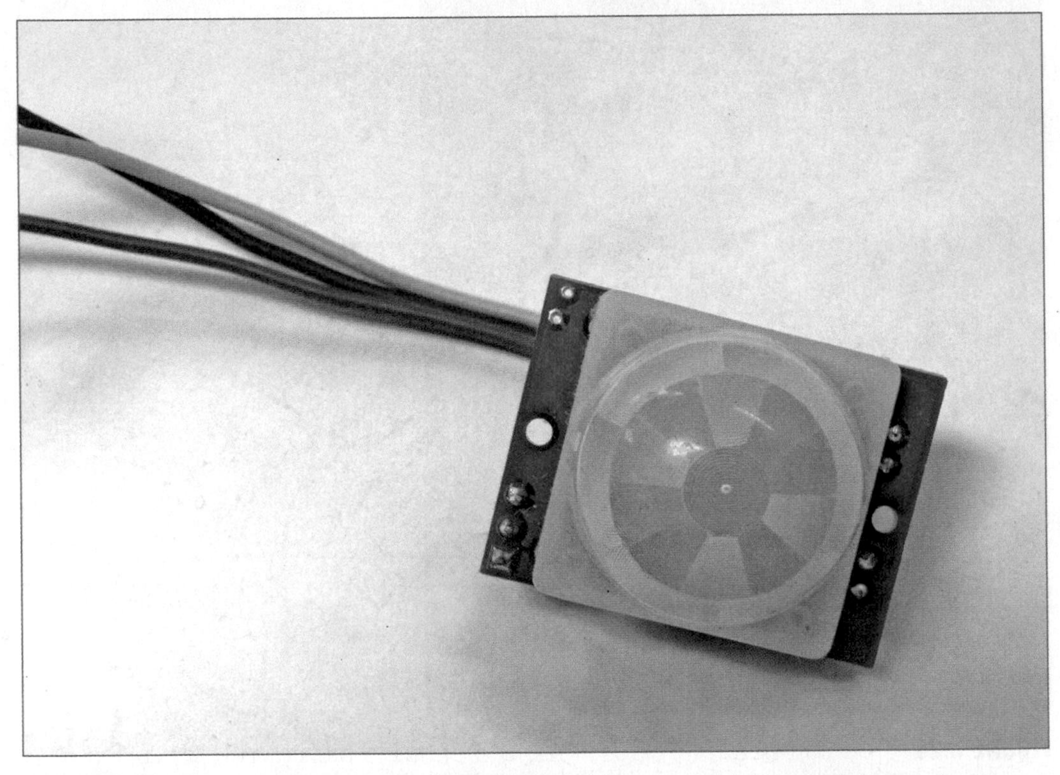

FIGURE 2.22 The passive infrared (PIR) sensor is a staple in professional security systems—so why not use it ourselves?

Called a PIR (passive infrared), the sensor detects subtle variations of infrared light in the area to determine whether someone or something has entered the sensing area. When it detects something, the PIR sends a signal to the Arduino.

Wiring Up the PIR and Buzzer

You need to get a PIR, which you can buy from Adafruit (P/N 189) for $10. It consists of a plastic bulb that shields the IR emitter and receiver. The circuit board has three terminals: One that takes power from the Arduino's 5V port, one that sends data to port 7, and one that goes to ground. Connect the PIR and buzzer as shown in Figure 2.23, and you're finished! The PIR senses in a 120-degree cone, about 20 feet long. Point the PIR toward the door you want to secure, and anyone coming through it or passing through the invisible cone of infrared light will set off the buzzer!

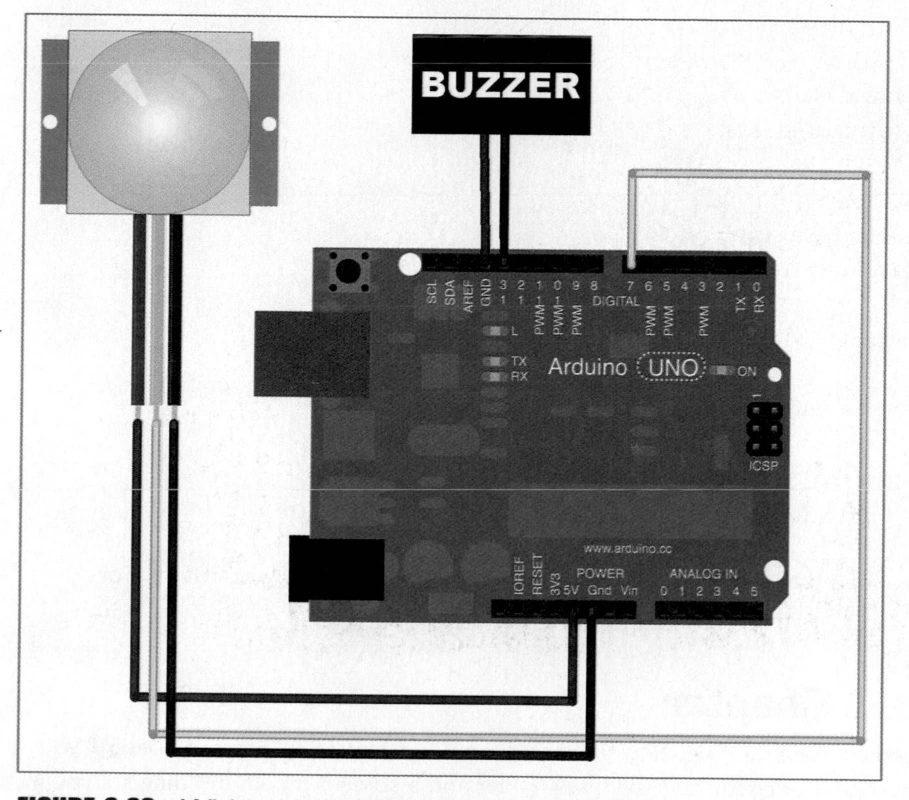

FIGURE 2.23 Wiring up the PIR is extremely simple!

Infrared Detector Code

Upload the following code to your Arduino to program your PIR alarm:

```
#define buzzerPin 13 // Pin 13 controls your LEDs

#define pirPin 4 // Pin 4 receives motion sensor data


int val = 0; // Sets a default for your motion sensor
```

```
void setup()
{
// Defines the buzzer and PIR as being input or output
  pinMode(buzzerPin, OUTPUT);
  pinMode(pirPin, INPUT);
     Serial.begin(115200);
}

void loop()
{
// The loop watches for the PIR to be triggered, then sets
// off the alarm
   val = digitalRead(pirPin);

   if (val == HIGH) {
     digitalWrite(buzzerPin, HIGH);
       Serial.println(val);

     delay(200); // alarm duration in milliseconds

   }
   else {
     digitalWrite(buzzerPin, LOW);
     }
   }
```

You can download the infrared detector code at https://github.com/n1/Arduino-For-Beginners.

The Next Chapter

You've mastered breadboarding. Next up is soldering! In Chapter 3, "How to Solder," you learn how to stick your circuit together with solder and a soldering iron. In doing so, you'll enhance a coffee table with a controllable light strip that displays groovy lighting effects.

How to Solder

Solder is an easily meltable alloy of lead and tin used to connect electronic components. Not only does it attach the part to the circuit board, it conducts electricity just like a wire does, allowing the circuit to function as you intended (see Figure 3.1).

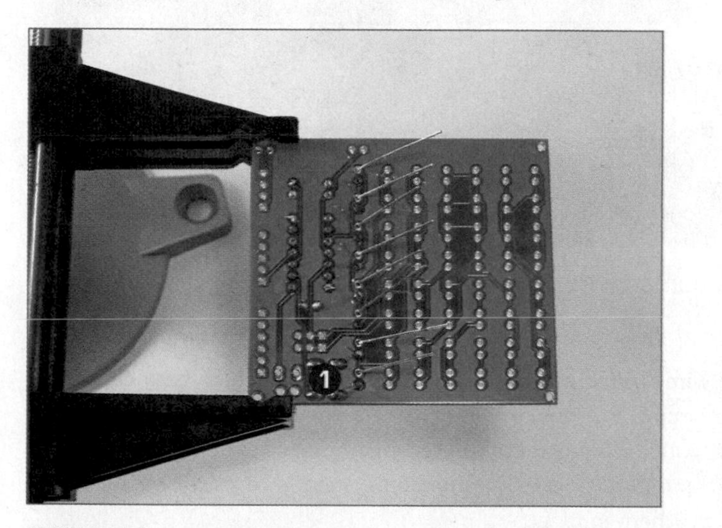

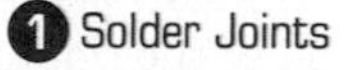

FIGURE 3.1 Soldering can look pretty intimidating, but doing it is actually easy. Credit: Wayne and Layne.

Most electronic kits come with a printed circuit board (PCB), shown in Figure 3.2, to which you solder the components. Typically, a board consists of a sheet of laminate drilled with numerous holes and screen printed with instructions.

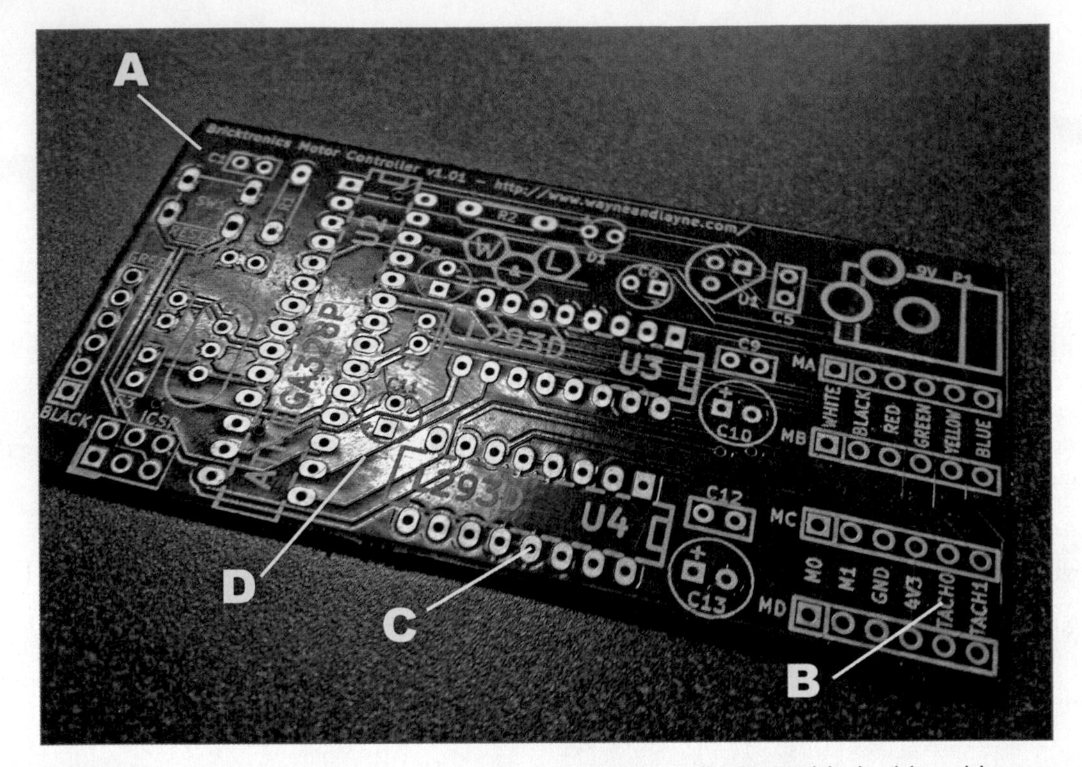

FIGURE 3.2 The typical PCB consists of a laminate plate studded with solder pads and wire traces.

The laminate is embedded with wires called *traces* (visible in Figure 3.2), which connect all the components together into a circuit. But how do you connect the components to the traces? If you look at the photo, you'll see tiny metal plates around each hole. These are *solder pads*. When you want to attach a part, you slide the component's wires (also called leads) through the hole and solder them in place.

Use the following key to identify the various carts of the circuit board shown in Figure 3.2:

A—Laminate board

B—Screen printing

C—Solder pads

D—Traces

The rest of this chapter guides you through learning to solder.

SOLDERING SAFETY

Not surprisingly, you can hurt yourself while soldering. Keep the following tips in mind:

- The soldering iron is hot; the tip can be upward of 600 degrees! It can burn you and also start fires if you're careless.
- Use eye protection when snipping leads. When you clip the excess wire off of electronic components, sometimes these leads fly off at high speeds. Putting on a pair of goggles to protect your eyes is not a bad idea, even if you wear regular glasses—sometimes, the projectile ricochets in from the side!
- Solder fumes are toxic. Make sure to have plenty of ventilation or even invest in a fume extractor (see "Fans or Fume Extractors," later in the chapter) to keep your air clean.
- Solder is lead. Lead is toxic. After you're finished with your project—or even when you take a break in the middle of it—be sure to wash your hands thoroughly. When you're working, be aware of the fact that your hands are likely to be toxic and don't touch your face with them.

Gathering Soldering Supplies

Not surprisingly, you need a soldering iron to solder. Some people might not realize that you need a bunch of other stuff as well! Following are some suggestions for equipment to buy.

Picking a Soldering Iron

Obviously, you need a soldering iron, but which one? As with anything, the cost ranges from inexpensive to pricy. A base model "pen style" iron (see Figure 3.3) typically consists of the soldering wand with a heat-up tip on one end and the electrical cord on the other. You can't adjust temperature or much of anything else. If you want to turn it off, you just unplug it.

You can find a decent, inexpensive soldering iron at www.adafruit.com/products/180.

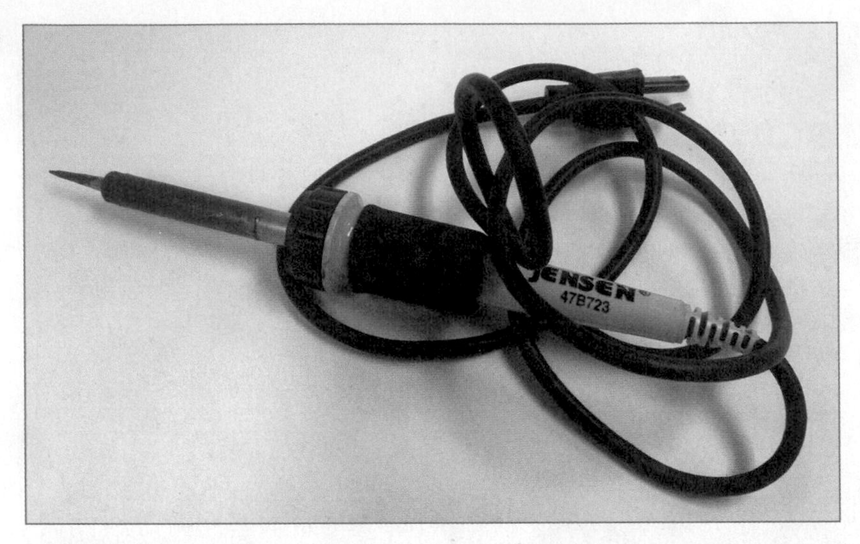

FIGURE 3.3 A pen-style soldering iron is a great choice for a beginning tinkerer.

A more complicated model, like the Weller WES51 shown in Figure 3.4, has more features. The Weller includes a soldering iron stand so the hot tip doesn't burn anything inadvertently. The stand also has a sponge for cleaning your tip, which is critical to maintaining the correct temperature.

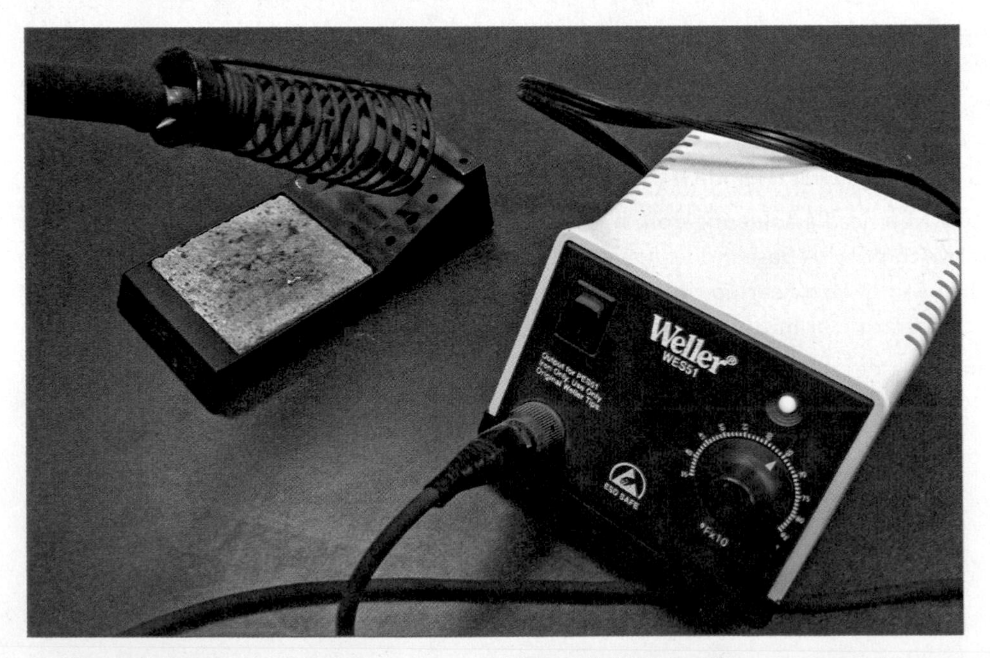

FIGURE 3.4 This relatively expensive Weller WES51 offers several features the pen irons lack.

More impressively, the expensive models have a more robust power supply. The Weller shown in Figure 3.4 enables you to dial in exactly how hot you want the iron to be, has a power switch, and even has an LED indicator telling you when the iron is at its designated temperature. The WES51 retails for about $100 more than the basic pen iron, but trust me, you'll be able to tell the difference. The greatest hindrance to learning how to solder is using a poor-quality iron.

You can buy the WES51 from Amazon: www.amazon.com/Weller-WES51-Analog-Soldering-Station/dp/B000BRC2XU/. (It's also cheaper than list price—score!)

TINNING YOUR TIP

No matter what iron you get, it will have a tip on it—this is the part that heats up. Soldering iron tips get easily corroded, which inhibits their capability to get hot. To keep your tip as pristine as possible, tin the tip after you're finished with it for the day (see Figure 3.5). Tinning means coating the tip in melted solder, and this protects the tip from corrosion. You'll also want to tin the tip periodically while you're actually soldering.

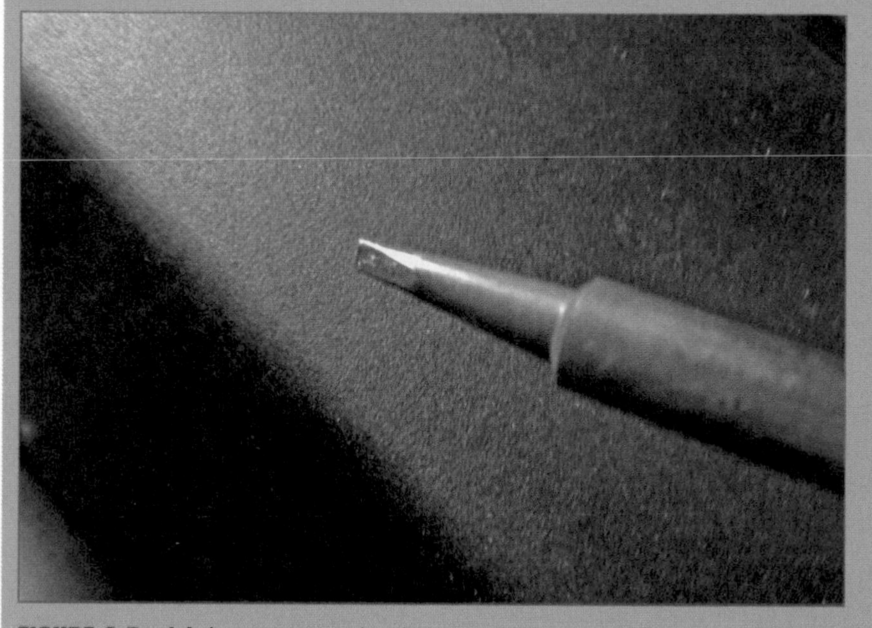

FIGURE 3.5 Make sure to tin the tip of your iron before, during, and after you solder.

Choosing a Solder

Although it's perhaps obvious that you'll need solder, choosing it is not quite so simple because several different types are available, as shown in Figure 3.6.

FIGURE 3.6 Many different gauges and alloys of solder are available. Make sure you choose the one that works best for your project.

Let's go over the various types:

- **Lead or lead-free**—Actually an alloy of tin and lead, lead is the most common type of solder. You can buy it in a variety of gauges (0.031"/0.8mm is a common one) and alloys (63/37 and 60/40 are typical) depending on your soldering needs and personal preference. Most tinkerers agree that lead makes the best solder; however, ecological laws and concerns over lead poisoning have caused manufacturers and hobbyists to turn to solder made without lead.
  Lead-free solder skips the lead in exchange for a cocktail of antimony, zinc, silver, and other ingredients that vary from product to product. A lot of makers don't like lead-free solder because it has a higher melting point than lead, it doesn't flow as readily, gives you messier solder joints, and can conceivably corrode over long periods of time. The recommendation is to stay with lead solder while you're learning; just wash your hands afterward!
- **Flux-core or solid-core**—Most solder comes with flux inside. This is used to chemically clean the surfaces, which strengthens the mechanical connection between solder and electronics and optimizes conductivity. The most popular type of flux is rosin, which is purified pine sap. Rosin-flux solders put out a lot of smoke, however, and the smoke can cause minor health problems, such as respiratory irritation and asthma-like breathing

difficulties. However, by using a fume extractor (see "Fans or Fume Extractors," later in the chapter) or just having plenty of ventilation, you can avoid taking in too much smoke. Solid-core solder doesn't contain flux; hardly anyone uses it anymore except for stained glass artisans who don't want flux stains on their creations.

- **Or just buy this one**—I know this is a lot of information to take in, so allow me to suggest the 0.31", rosin-core 60/40 lead solder, which is a nice all-purpose solder that I use for all of my projects. You can buy it at www.adafruit.com/products/145.

Getting the Other Things You Need

You have a soldering iron and solder—now what? Let's go over some additional accessories that you can use to make your soldering experience easy and successful.

Desktop Vises

Called a "third hand," the rig shown in Figure 3.7 is used to hold a circuit board steady with alligator clips while you solder. Some models also come with a magnifying glass, which can be helpful if you're doing some challenging solder joints and need a closer look.

FIGURE 3.7 The third hand holds your soldering project still while you work on it.

You can get a decent one at www.makershed.com/product_p/mkhh1.htm.

Panavise (panavise.com) manufactures small desk vises. Its Panavise Jr. Model 201 (see Figure 3.8) is extremely popular among tinkerers as a way of holding PCBs steady during soldering. It's essentially a small vise that you can attach to your workbench with bolts, and that holds the PCB securely at any angle.

You can pick one of these up at www.makershed.com/Panavise_Jr_Model_201_p/mkpv01.htm.

FIGURE 3.8 A Panavise Jr. is another great way of holding your circuit board steady.

Cutters and Strippers

You'll definitely need wire cutters and strippers (see Figure 3.9) for trimming wires to length and stripping off insulation. Some models (like this inexpensive one: www.adafruit.com/products/147) combine both strippers and cutters, but I prefer having separate tools.

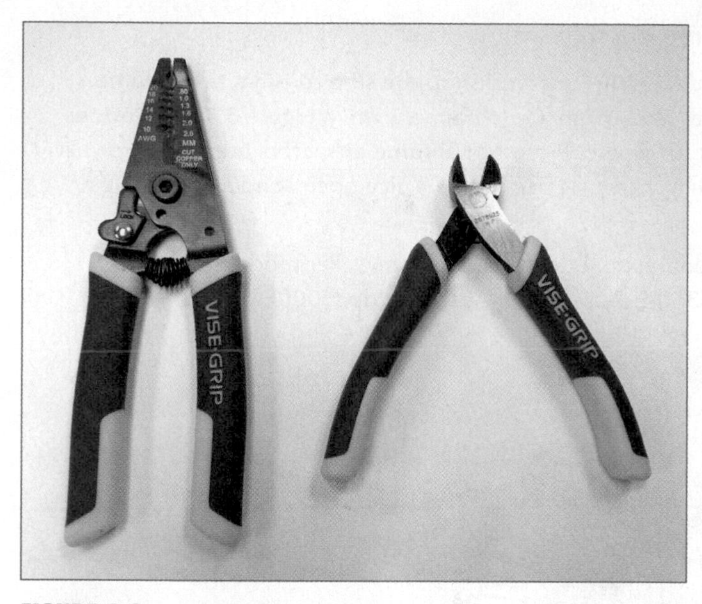

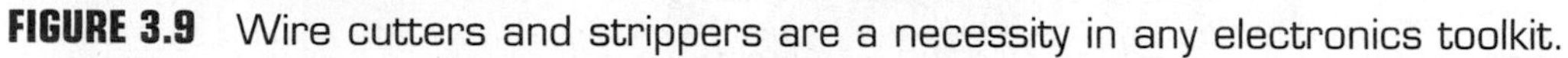
FIGURE 3.9 Wire cutters and strippers are a necessity in any electronics toolkit.

Needle-Nose Pliers and Hemostats

You might also need needle-nose pliers and hemostat medical clamps (see Figure 3.10) to grab small items—electronics have a lot of tiny objects! Adafruit offers some inexpensive tweezers (www.adafruit.com/products/421) with a non-conductive coating that helps minimize the chance of accidentally statically shocking your components.

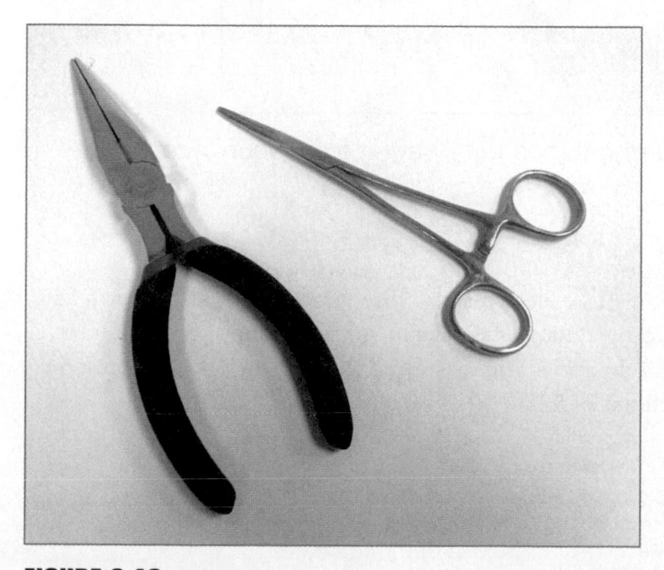

FIGURE 3.10 Need to grab or hold a small part? Needle-nose pliers or a hemostat is just what you need.

Fans or Fume Extractors

If you don't have very good ventilation in your workshop, be sure to use a fan or fume extractor to blow the rosin fumes away from you. A fan, shown in Figure 3.11, is obvious, and you can buy one just about anywhere. Professional fume extractors are more expensive and include cooler features such as carbon-fiber filters. A nice open window is mainly what you need, however!

If you are interested in professional fume extractors, see this Weller model: www.amazon.com/Weller-WSA350-Bench-Smoke-Absorber/dp/B000EM74SK/).

FIGURE 3.11 A fan carries soldering fumes safely away from your face.

ESD Protection

One threat to your electronic components is electro-static discharge (ESD), also known as your garden-variety static shock. If you get a lot of shocks in your workshop, or if you just don't want to take any chances, wear an anti-static wristband (see Figure 3.12) or work on an anti-static mat to minimize the threat of ESD.

FIGURE 3.12 Wearing an anti-static strap protects your project from electrostatic discharge.

Belkin makes a good and inexpensive wristband: www.amazon.com/Belkin-Anti-Static-Wrist-Adjustable-Grounding/dp/B00004Z5D1/.

Solder Stand and Sponge

Finally, if you have a pen-style iron, you might want to buy a separate stand to hold your iron (see Figure 3.13), and you'll definitely need a sponge to keep the tip clean. Adafruit offers a nice stand-and-sponge combo (www.adafruit.com/products/1154) that also includes a solder dispenser.

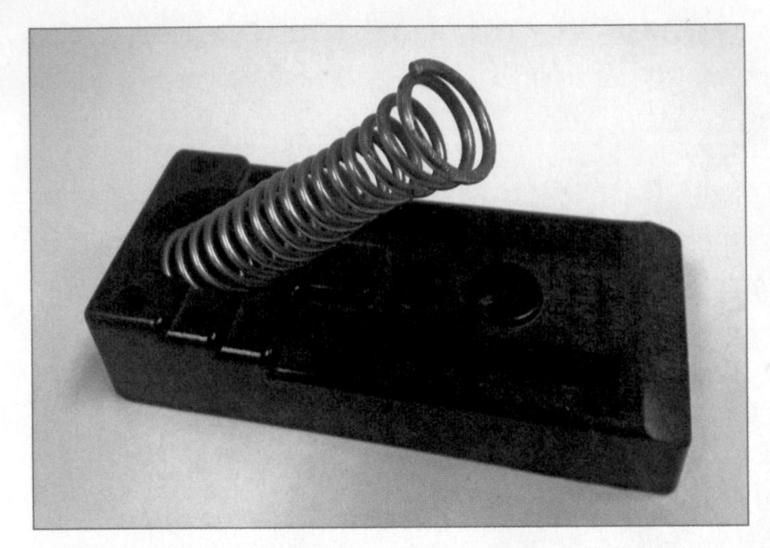

FIGURE 3.13 A soldering iron is hot, so you should keep it safely off the table.

Soldering

Now that you have your equipment, it's time to solder! Here's how:

1. Prepare your soldering equipment and work area (see Figure 3.14). Make sure you have plenty of space in which to work and that your wire cutters, sponge, and other tools are ready to go. Plug in your iron, and if it's the type that needs to be turned on, turn it on. If your iron has an adjustable temperature control, set it to 700°F/370°C for tin-lead solder and 750°F/400°C for lead-free solder.
2. Your iron heats up, and if you have one with a readout, it will tell you when it is hot. If you just have a basic model iron, you must test the tip to ensure that it's ready to solder. Touch the tip to your wet sponge; if the iron is ready, a tiny wisp of steam will hiss out.
3. After the soldering iron is hot, melt some solder and coat the iron's tip with it. This is called tinning, and it helps conduct heat easier and thereby speeds up your soldering.

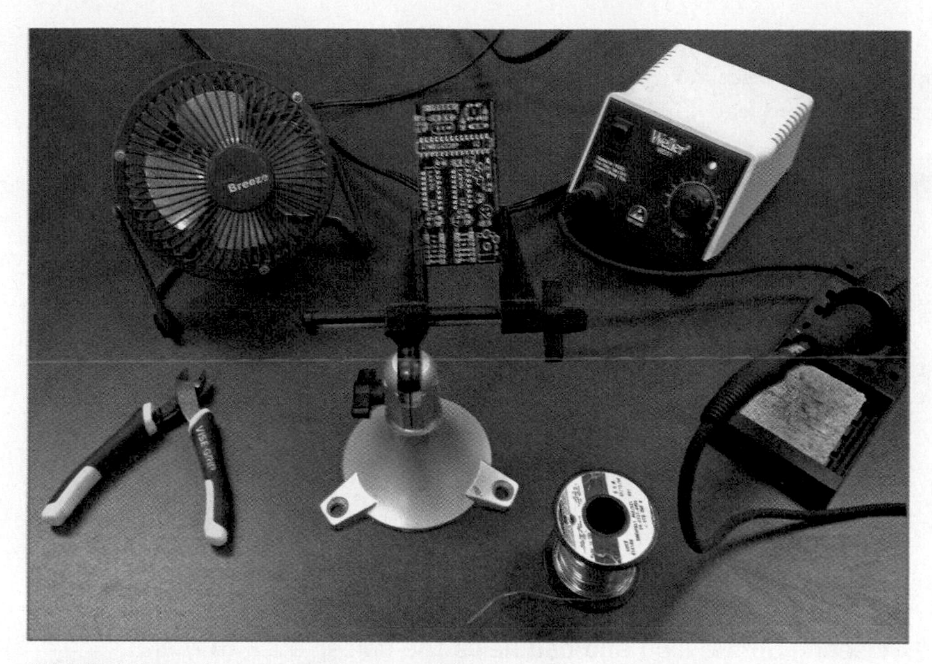

FIGURE 3.14 Have everything ready to go? Let's get started!

4. When you're ready, go to step 1 of your instructions, assuming you're working from a kit, which is how most beginners learn. Kit instructions typically guide you through the placement of each component in turn. Slide the component's leads through the holes in the solder pads. Flip over the board and bend back the leads (as shown in Figure 3.15) so the component doesn't fall back out.

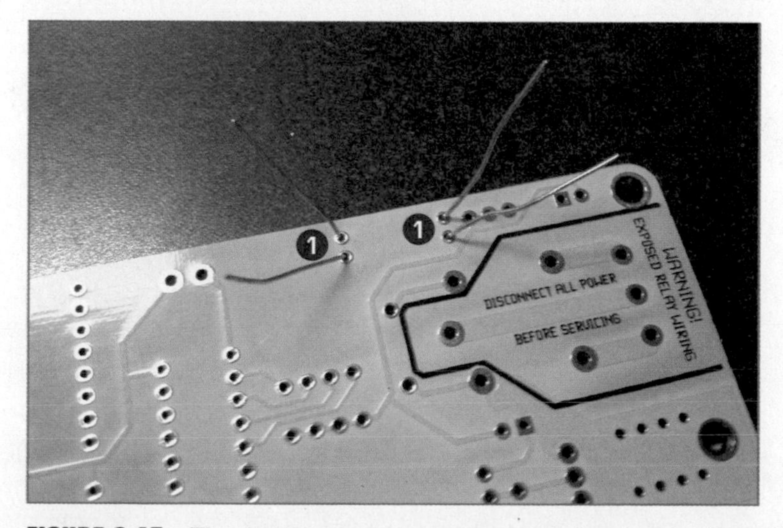

1 Bend the component leads just enough to prevent them from falling back out.

FIGURE 3.15 The best way to keep your components from falling out when you turn the PCB over is to bend back the leads.

5. Touch the tip of the hot soldering iron to both the circuit board's pad and the lead of the component, as you see in Figure 3.16. Hold it there for a couple of seconds. This warms up the pad and lead and helps the solder stick to them.

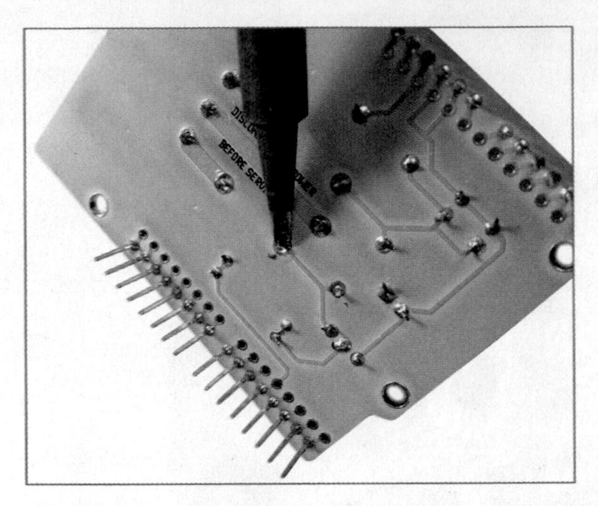

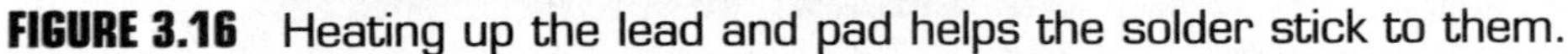

FIGURE 3.16 Heating up the lead and pad helps the solder stick to them.

6. Hold a piece of solder with your other hand and touch the end to the pad and lead while the iron is still touching them (see Figure 3.17). The solder should melt immediately and flow into the hole, sticking everything together. Remove the iron and you're finished! Cooling takes barely a second, so you don't have to wait before moving on to the next step.

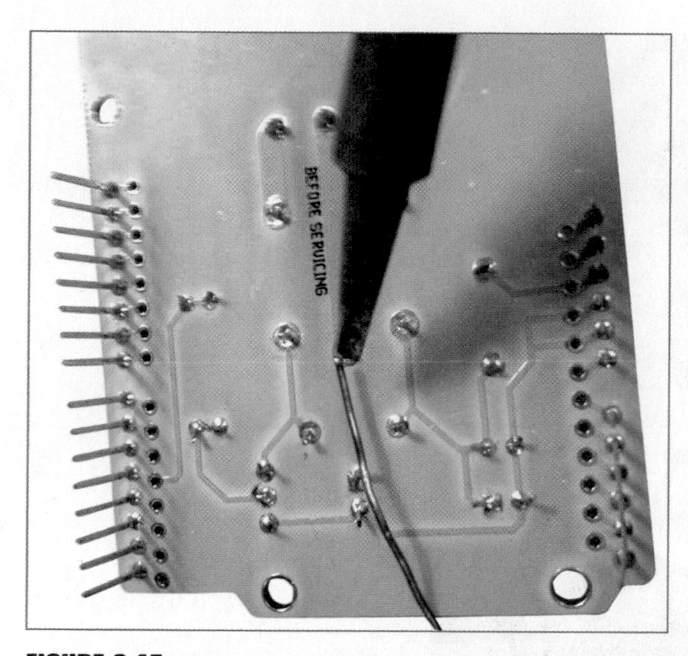

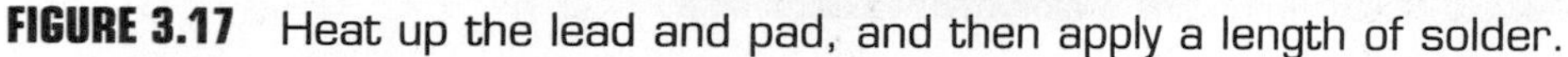
FIGURE 3.17 Heat up the lead and pad, and then apply a length of solder.

7. Examine the solder bead. It should cover the entire pad and there should be enough solder that it forms a small bump. If the solder bead is flat against the pad or if you can see extra pad sticking out from under the solder, then you probably didn't use enough solder and you might run into problems. Conversely, if you used too much solder, the bead might touch more than one pad and cause the circuitry to not function as intended. Either way, you should probably desolder (see the next section) and redo your work.
8. If the solder looks good, clip off the excess leads; you won't need them. See Figure 3.18. You can then move on to the next component.

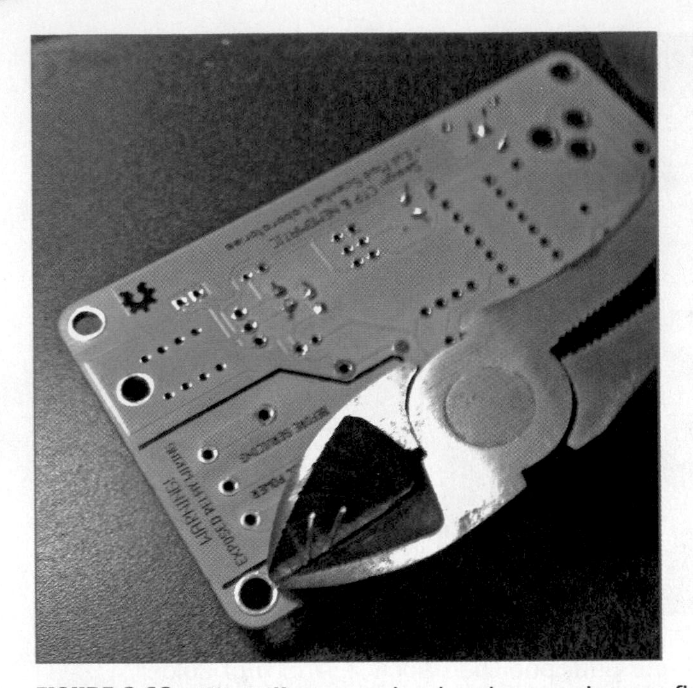

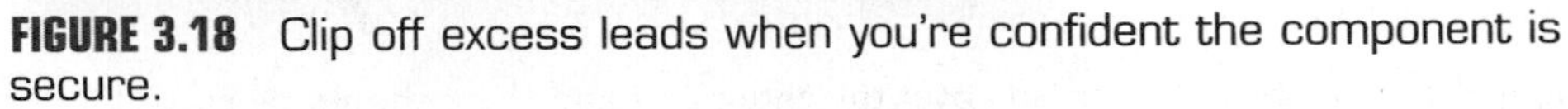

FIGURE 3.18 Clip off excess leads when you're confident the component is secure.

Desoldering

Sometimes your soldering effort results in a bad connection, as you can see in Figure 3.19. Maybe you didn't use enough solder, or maybe you used too much solder and the glop of metal covers more than one pad. Sometimes you accidentally attach the wrong part or solder it in backward. In these cases, you must remove the solder and redo your work.

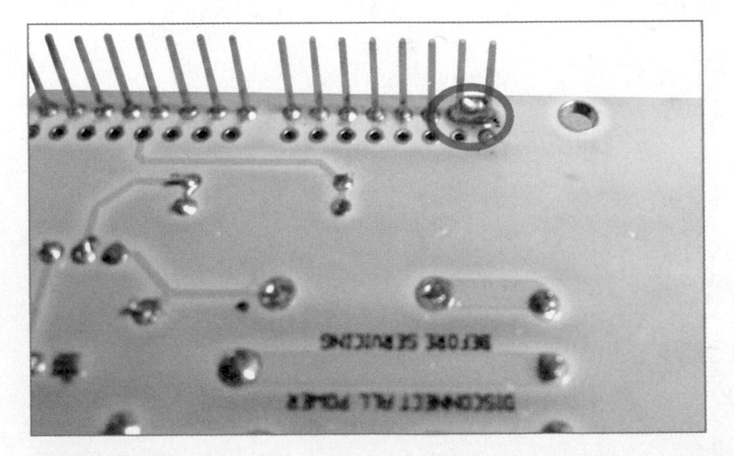

FIGURE 3.19 See the two pins stuck together? That's a bad solder joint.

Desoldering uses a number of tools (see Figure 3.20) to help remove melted solder. These consist of a desoldering bulb and braid, as well as a solder sucker. Ultimately you'll need to come up with the method that works for you, but for the record, I like the solder sucker the best!

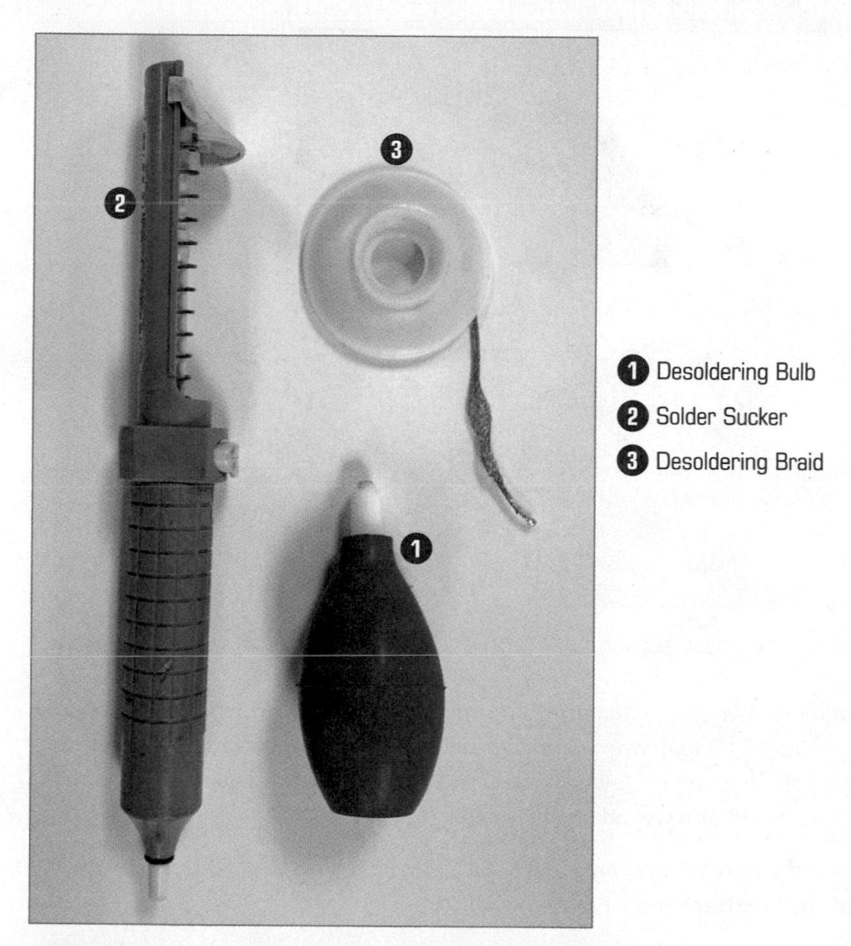

FIGURE 3.20 These are the tools you need to desolder.

To desolder, you need the following tools:

- **Desoldering bulb**—This is a hollow rubber bulb with a nozzle. To desolder, hold the nozzle of the bulb to the solder bead and melt the bead with your iron. To suck up solder, you simply squeeze on the bulb while holding the nozzle up to the melted solder. You stop squeezing and the solder is vacuumed into the bulb. You can buy a desoldering bulb at www.radioshack.com/product/index.jsp?productId=2062742.
- **Solder sucker**—This is a spring-loaded version of the bulb and comes with a plunger and button. When you think you'll need to desolder, you press down the plunger; when

it clicks, you know it's ready to go. Hold the sucker's nozzle next to the bead of molten solder and press the button. The spring releases and the plunger pops back, creating a vacuum that sucks the solder away from the circuit board (see Figure 3.21). You can buy one at www.adafruit.com/products/148.

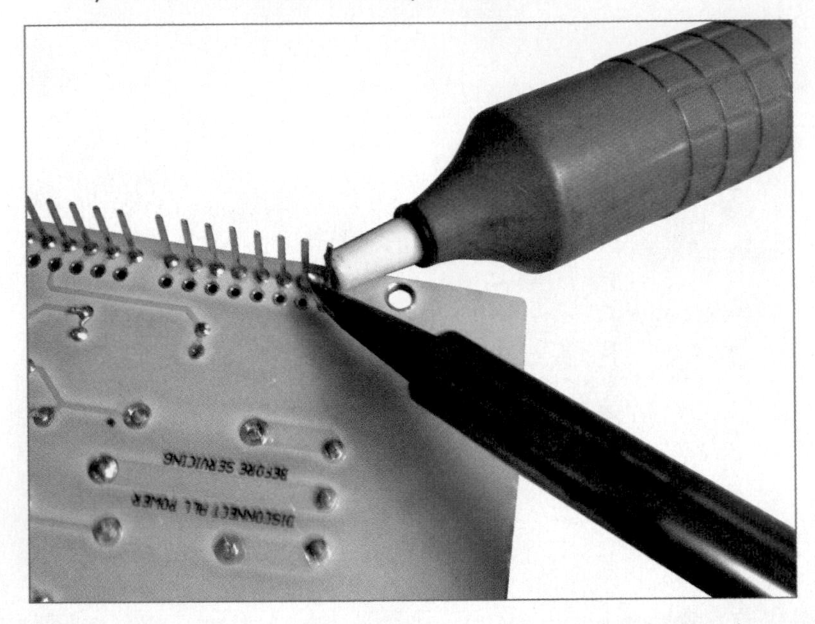

FIGURE 3.21 Heat up the bad solder joint and suck up the molten metal!

- **Desoldering braid**—Rather than attempting to suck up the solder, why not sop it up like a puddle of spilled milk? Desoldering braid is loosely braided wire thread, and when it touches melted solder, the solder flows up the braid and away from your project. You can buy desoldering braid at www.adafruit.com/products/149.

When you've desoldered a component, examine it carefully to ensure that no large glops of solder are on it, and then reattach it.

SOLDERING TIPS

Here are some suggestions that can help your soldering experience go smoother:

- **Better too little than too much**—You don't need a lot of solder to make a good joint. In fact, too much might cause two solder pads to connect when they shouldn't.
- **Tin the tip**—Periodically re-tin the tip of your soldering iron to ensure enough heat reaches the solder.
- **Bend the leads**—When you insert a component, bend the leads to ensure the part doesn't fall out.
- **Solder one lead at a time**—If you're worried about a part being crooked, solder just one of its leads, then heat up the solder again and adjust the fit; when it's straight, let the solder cool and solder the remaining lead(s).
- **Heat the pad and leads, not the solder**—One common beginners' mistake is to melt the solder and smear it all over the leads. Do it the opposite way: Heat the pad and leads, and then apply a length of solder and let it melt.
- **Keep your tip clean**—Clean the tip periodically during the soldering process by wiping it off on the soldering iron's sponge. Re-tin, and then continue soldering.
- **Tin your tip before storing**—Store the soldering iron with a tinned tip; this helps keep the tip from corroding.

Cleanup

You're finished! Congratulations on learning a new skill. Now it's time to put away your tools and clean up your work area. The following are suggestions on what to do:

1. Tin the tip of your iron. You learned how to do this earlier in the chapter. Covering the tip in solder helps protect it against corrosion when not in use.
2. Unplug your tools and put them away.
3. Clean your work area. There is likely to be some tiny specks of toxic lead as well as clipped leads on the workbench and the floor by your chair. Use a broom or vacuum on the floor and wipe down the table with a typical multi-surface spray cleaner.
4. The final step should be to wash your hands one last time to make absolutely certain all the lead is cleaned up.

Project: LED Strip Coffee Table

In the next project, you learn how to add a cool programmable lighting strip to an ordinary coffee table, spicing up your next coffee klatch! The strip consists of a metal foil strip studded with LEDs and microchips, and you can control each LED individually, with brightness and color set by an Arduino program (see Figure 3.22). You'll be able to toggle through various cool lighting effects with a button, enabling you to find the one that you want within seconds.

FIGURE 3.22 Your coffee table will light up your next social function—literally.

PARTS LIST

You won't need many parts to build this project:

- Arduino Uno
- Wall wart for the Uno, which also powers the LED strip
- Digital LED light strip (Adafruit P/N 306: Get however many meters you think you'll need.)
- Coffee table
- Jumpers
- A button (I used a U811SHZGE pushbutton from Digi-Key.)
- Potentiometer (Adafruit P/N 562)
- Zip ties (optional)
- Hot glue gun and glue (optional)

Preparing the Light Strip

Digital Red Green Blue (RGB) LED strips consist of a strip of metal embedded with tiny microchips and LEDs. There are 32 LEDs per meter, each of which can be addressed individually, with brightness and color fully controlled by the Arduino (see Figure 3.23).

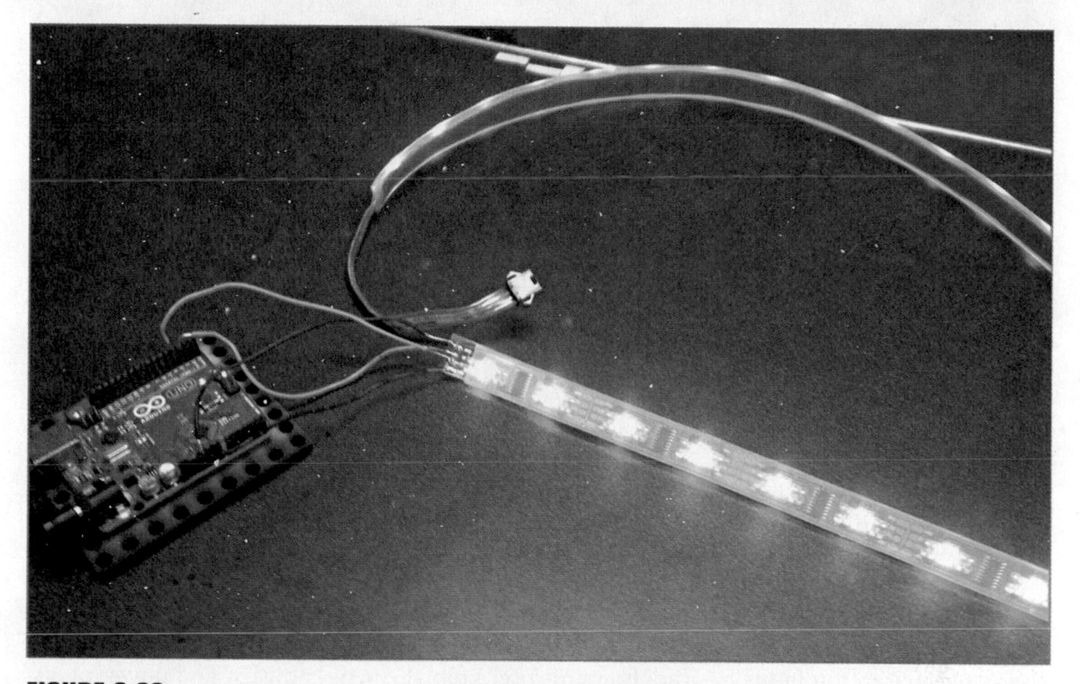

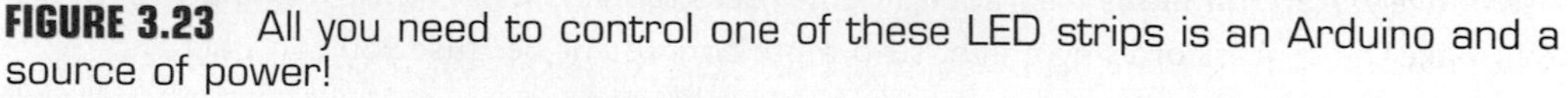

FIGURE 3.23 All you need to control one of these LED strips is an Arduino and a source of power!

One of the first things you might notice is that the metal strip is covered in a clear plastic sleeve that protects it from moisture. The strip doesn't need the sleeve to operate, but don't remove it unless you absolutely must. Also, the sleeve is really difficult to get back on the strip after it's removed!

You buy the strip in five-meter reels but you can cut it yourself into lengths as small as 2.5". Does it sound scary to potentially damage a light strip that costs $30 a meter? It should. However, the manufacturers thoughtfully created the strip so that it can be cut along certain cutlines. The safe cutting line is flanked by solder pads, as shown in Figure 3.24.

To prepare the light strip for the project, follow these steps:

1. To cut the strip, find the nearest cutline—you can't miss them, they're every 2.5 inches! Simply cut down the line with a sturdy pair of scissors. It doesn't have to be a surgical cut; the shoddy job I did in Figure 3.24 worked perfectly fine.

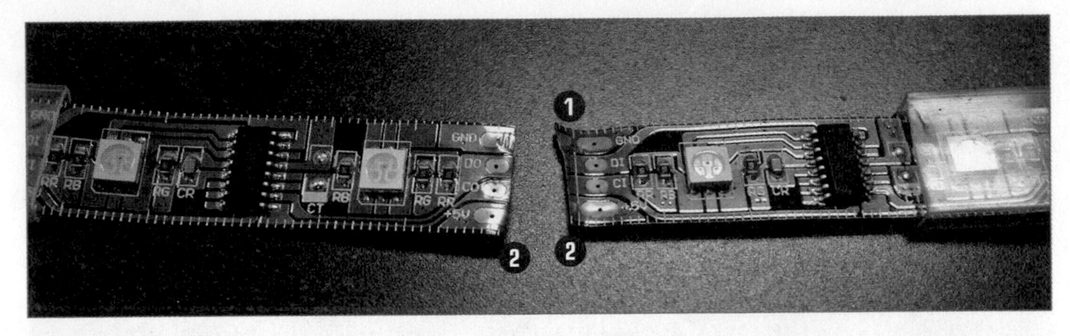

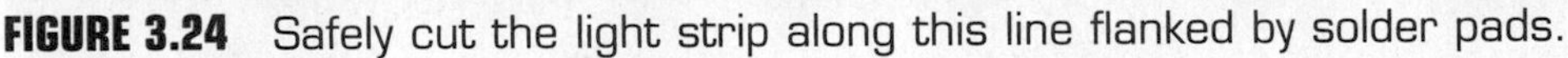
FIGURE 3.24 Safely cut the light strip along this line flanked by solder pads.

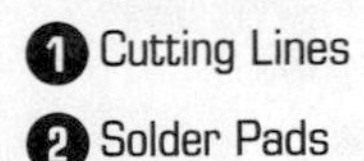
1 Cutting Lines

2 Solder Pads

2. Connect the separated strips back into a long one, to enable you to more effectively wrap it around the table. Grab two lengths and lay them next to each other. Pull back the plastic sleeve an inch or two to expose the solder pads.
3. Add a dab of solder to each pad, and then tin the end of your jumper—you probably need less than a quarter-inch of exposed wire—and solder it to the pad, as shown in Figure 3.25. Some people also use hot glue on the exposed contacts to ensure the connection remains solid.
4. Connect the remaining contacts the same way.

Alternatively, Adafruit Industries (adafruit.com) sells replacement end caps and power plugs that make the process of using a light strip more convenient because you can plug and unplug the lengths at will, which saves you some soldering time.

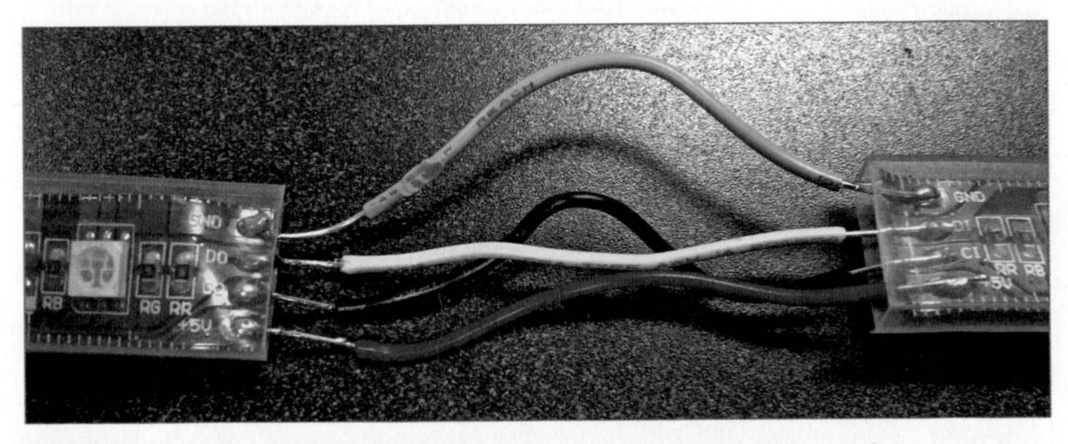

FIGURE 3.25 Connect each solder pad to its mate on the other strip.

Attaching the Light Strip to the Table

Every model of coffee table differs from the next, so needless to say, you'll have to adapt certain parts of this project to the unique needs of your coffee table. Specifically, the LED strip must be the same length as the circumference of the table.

1. Measure the coffee table to find out how much of the LED strip you'll need. Be sure to be cut the strips smaller than the length, so you don't have the strip sticking out too far.
2. Cut the LED strip into the right-sized segments. My table measured 42" by 20" so I cut my three meters of LED strip into two 42" segments and two 17" segments. (You might not actually need to cut the strip apart, depending on your coffee table!)
3. Wire up the segments—you learned how to do this earlier in this chapter. See "Preparing the Light Strip."
4. Attach the light strip to the table, which you can do in any number of ways, including hot glue or zip ties. The method you choose depends on your table.
5. Wire up the Arduino, button, and potentiometer, as shown in Figure 3.26.

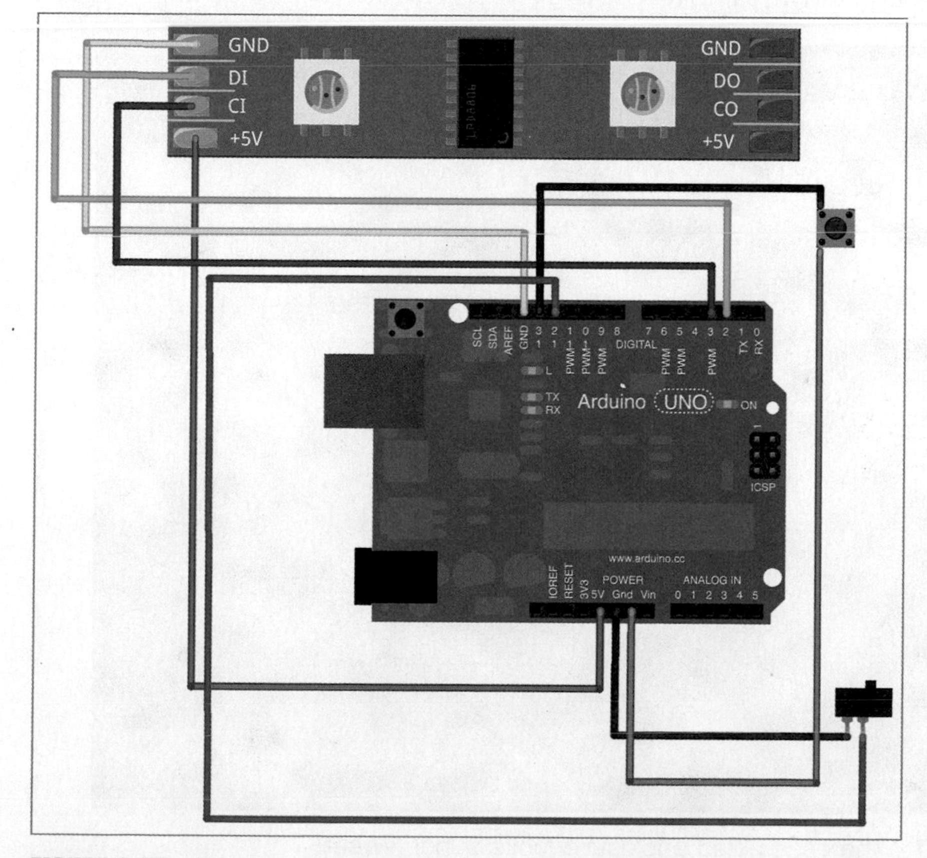

FIGURE 3.26 Wire up the light strip as you see here.

- Connect the light strip to the Arduino; GND on the strip goes to GND on the Arduino. The pad marked 5V connects to 5V on the Arduino. DI connects to pin 2 and CI connects to pin 3.
- The positive terminal of the button plugs into RESET on the Arduino and the negative connects to GND.
- The center terminal of the potentiometer connects to A0 on the Arduino, while one of the other two terminals plugs into GND and the other into 5V—it doesn't matter which goes where.

Building the Enclosure

For this project, you build the enclosure out of MicroRAX. This is a line of fairly cheap but extremely durable aluminum beams used to build computer chassis and robots. You can buy it at MicroRAX.com and Sparkfun.com. You can buy whole kits, individual beams, and longer beams (up to eight feet) that you can saw down to smaller sizes. After you have the MicroRAX framework, you add acrylic panels to serve as the sides of the box (see Figure 3.27).

FIGURE 3.27 The assembled enclosure looks super sweet!

PARTS LIST

Use the following parts to build your enclosure:

- **MicroRAX beams** (available from MicroRAX.com)
  Four 40mm beams
  Four 100mm beams
  Four 160mm beams
- **Eight tri-corner braces** (available from MicroRAX.com)
- **Six 90-degree joining brackets** (available from MicroRAX.com)
- **3mm screws** (available from MicroRAX.com)
- **Acrylic or wood panels** for the sides of the box
- **Standoffs**, 3/8-inch, Sparkfun P/N 10461
- **Machine screws**, #4-40 × 1"

The assembly is super simple!

1. Connect the beams to form a box, using the 90-degree joining brackets and 3mm screws. You'll be adding pieces of acrylic or wood to form the top, bottom, and sides.
2. Get your beams together. I used four each of 40mm, 100mm, and 160mm beams (see Figure 3.28) but make yours however you want.

FIGURE 3.28 MicroRAX beams come in a variety of pre-cut lengths; I used 40mm, 100mm, and 160mm beams to build my enclosure.

3. Use corner braces (see Figure 3.29) to start building the framework of the box.

FIGURE 3.29 MicroRAX corner braces allow you to easily connect multiple beams.

4. Cut out some panels for the sides. I laser-cut mine out of acrylic (see Figure 3.30) but this isn't absolutely necessary; you could cut them out of wood with a regular saw just fine. Make the panels the same size as the beams but add 4mm–6mm to the length and width of the panel if you're using 2.5mm or thinner stock and 2mm–3mm if you're using 5mm stock, the maximum thickness. If they're a bit loose and rattle, stick a piece of rubber band in there to pad it.

FIGURE 3.30 Slide the panels into the beams' grooves.

5. While you're building the frame, be sure to add angle braces (shown in Figure 3.31) to connect the enclosure to the coffee table. Note that I made the bottom panel out of wood to make it easier to connect the Arduino to the enclosure. Finish adding the other panels and beams, leaving the top unsecured until you add the electronics.

FIGURE 3.31 The enclosure begins to take shape.

6. Attach your Arduino to the base. If you're planning ahead, you can laser cut the screw-holes. If you forgot, like me, then place the Arduino on the base and use a pen to mark where the holes on the Arduino's PCB are located. Drill holes. Then, thread your #4-40 machine screws through the base and add the 3/8-inch standoffs. Place the Arduino on the screws and then tighten the nuts. When you're done, it should look just like Figure 3.32.

FIGURE 3.32 The standoffs and screws keep the Arduino positioned correctly.

7. After the guts are in the enclosure, close up the top and lightly secure it with one of the MicroRAX screws. You won't want to secure it fully until you're done. Figure 3.33 shows you how the final enclosure looks.

FIGURE 3.33 The enclosure attached to the underside of the coffee table.

Controlling the LED Strip

You use the potentiometer and the button to control which of the eight effects the LED strip displays. This is how it works:

1. The potentiometer has been mapped to return a value of 1 to 8 depending on how it's turned. (I explain mapping in Chapter 5.)
2. When the sketch is launched, the Arduino takes that number and displays whichever effect is currently selected.
3. However, if you want to change the effect while the Arduino is running, you'll have to press the button, which resets the Arduino, to see the new effect. This is because the sketch is looping and doesn't recognize that the potentiometer has changed until you press reset.

LED Strip Code

Upload the following code to make your coffee table project come alive.

NOTE

Code Available for Download

You don't have to enter all of this code by hand. Simply go to https://github.com/n1/Arduino-For-Beginners to download the free code.

Uploading code to your Arduino is explained in Chapter 2, "Breadboarding," and Chapter 5, "Programming Arduino." Also, you'll need the LPD8806.h library (libraries are explained in Chapter 5), which can be downloaded from the following URL: https://github.com/adafruit/LPD8806/blob/master/LPD8806.h.

```
//This sketch is derived from Adafruit's LPD8806 example code
#include "LPD8806.h"
#include "SPI.h"

int pot1 = A1;
int dataPin = 2;
int clockPin = 3;
int toggleValue = 0;
int toggle = 0;

//the 96 refers to the number of LEDs on your strip. Change the number as needed.
LPD8806 strip = LPD8806(96, dataPin, clockPin);

void setup() {


    pinMode(pot1, INPUT);
    Serial.begin(9600);


  // Start up the LED strip
  strip.begin();

  // Update the strip, to start they are all 'off'
  strip.show();
```

```
}

void loop() {

toggle = analogRead(pot1);

int toggleStatus = map(toggle, 0, 1023, 0, 8);

    Serial.println(toggleStatus);


switch(toggleStatus) {

  case 0:
    // Clear strip data before start of next effect
    for (int i=0; i < strip.numPixels(); i++) {
      strip.setPixelColor(i, 0);
    }
  break;

  case 1:
    // Send a simple pixel chase in...
    colorChase(strip.Color(127,127,127), 20); // white
    colorChase(strip.Color(127,0,0), 20);      // red
    colorChase(strip.Color(127,127,0), 20);    // yellow
    colorChase(strip.Color(0,127,0), 20);      // green
    colorChase(strip.Color(0,127,127), 20);    // cyan
    colorChase(strip.Color(0,0,127), 20);      // blue
    colorChase(strip.Color(127,0,127), 20);    // magenta
  break;

  case 2:
    // Fill the entire strip with...
    colorWipe(strip.Color(127,0,0), 20);       // red
    colorWipe(strip.Color(0, 127,0), 20);      // green
    colorWipe(strip.Color(0,0,127), 20);       // blue
    colorWipe(strip.Color(0,0,0), 20);         // black
  break;

  case 3:
    // Color sparkles
```

```
      dither(strip.Color(0,127,127), 50);         // cyan, slow
      dither(strip.Color(0,0,0), 15);              // black, fast
      dither(strip.Color(127,0,127), 50);         // magenta, slow
      dither(strip.Color(0,0,0), 15);              // black, fast
      dither(strip.Color(127,127,0), 50);         // yellow, slow
      dither(strip.Color(0,0,0), 15);              // black, fast
    break;

    case 4:
      // Back-and-forth lights
      scanner(127,0,0, 30);           // red, slow
      scanner(0,0,127, 15);           // blue, fast
    break;

    case 5:
      // Wavy ripple effects
      wave(strip.Color(127,0,0), 4, 20);          // candy cane
      wave(strip.Color(0,0,100), 2, 40);          // icy
    break;

    case 6:
      // make a pretty rainbow cycle!
      rainbowCycle(0);  // make it go through the cycle fairly fast
    break;

    case 7:
      rainbowCycle(0);
    break;

    case 8:
    // Color sparkles
      dither(strip.Color(0,127,127), 50);         // cyan, slow
      dither(strip.Color(0,0,0), 15);              // black, fast
      dither(strip.Color(127,0,127), 50);         // magenta, slow
      dither(strip.Color(0,0,0), 15);              // black, fast
      dither(strip.Color(127,127,0), 50);         // yellow, slow
      dither(strip.Color(0,0,0), 15);              // black, fast
    break;

}
}
```

```
/* Helper functions */

//Input a value 0 to 384 to get a color value.
//The colours are a transition r - g - b - back to r

uint32_t Wheel(uint16_t WheelPos)
{
  byte r, g, b;
  switch(WheelPos / 128)
  {
    case 0:
      r = 127 - WheelPos % 128; // red down
      g = WheelPos % 128;       // green up
      b = 0;                    // blue off
      break;
    case 1:
      g = 127 - WheelPos % 128; // green down
      b = WheelPos % 128;       // blue up
      r = 0;                    // red off
      break;
    case 2:
      b = 127 - WheelPos % 128; // blue down
      r = WheelPos % 128;       // red up
      g = 0;                    // green off
      break;
  }
  return(strip.Color(r,g,b));
}

// Cycle through the color wheel, equally spaced around the belt
void rainbowCycle(uint8_t wait) {
  uint16_t i, j;

  for (j=0; j < 384 * 5; j++) {     // 5 cycles of all 384 colors in the wheel
    for (i=0; i < strip.numPixels(); i++) {
      strip.setPixelColor(i, Wheel(((i * 384 / strip.numPixels()) + j) % 384));
    }
    strip.show();   // write all the pixels out
    delay(wait);
  }
}
```

```
// Chase one dot down the full strip.
void colorChase(uint32_t c, uint8_t wait) {
  int i;

  // Start by turning all pixels off:
  for(i=0; i<strip.numPixels(); i++) strip.setPixelColor(i, 0);

  // Then display one pixel at a time:
  for(i=0; i<strip.numPixels(); i++) {
    strip.setPixelColor(i, c); // Set new pixel 'on'
    strip.show();              // Refresh LED states
    strip.setPixelColor(i, 0); // Erase pixel, but don't refresh!
    delay(wait);
  }

  strip.show(); // Refresh to turn off last pixel
}

// Fill the dots progressively along the strip.
void colorWipe(uint32_t c, uint8_t wait) {
  int i;

  for (i=0; i < strip.numPixels(); i++) {
      strip.setPixelColor(i, c);
      strip.show();
      delay(wait);
  }
}

// An "ordered dither" fills every pixel in a sequence that looks
// sparkly and almost random, but actually follows a specific order.
void dither(uint32_t c, uint8_t wait) {

  // Determine highest bit needed to represent pixel index
  int hiBit = 0;
  int n = strip.numPixels() - 1;
  for(int bit=1; bit < 0x8000; bit <<= 1) {
    if(n & bit) hiBit = bit;
  }

  int bit, reverse;
  for(int i=0; i<(hiBit << 1); i++) {
```

```
    // Reverse the bits in i to create ordered dither:
    reverse = 0;
    for(bit=1; bit <= hiBit; bit <<= 1) {
      reverse <<= 1;
      if(i & bit) reverse |= 1;
    }
    strip.setPixelColor(reverse, c);
    strip.show();
    delay(wait);
  }
  delay(250); // Hold image for 1/4 sec
}

// "Larson scanner" = Cylon/KITT bouncing light effect
void scanner(uint8_t r, uint8_t g, uint8_t b, uint8_t wait) {
  int i, j, pos, dir;

  pos = 0;
  dir = 1;

  for(i=0; i<((strip.numPixels()-1) * 8); i++) {
    // Draw 5 pixels centered on pos.  setPixelColor() will clip
    // any pixels off the ends of the strip, no worries there.
    // we'll make the colors dimmer at the edges for a nice pulse
    // look
    strip.setPixelColor(pos - 2, strip.Color(r/4, g/4, b/4));
    strip.setPixelColor(pos - 1, strip.Color(r/2, g/2, b/2));
    strip.setPixelColor(pos, strip.Color(r, g, b));
    strip.setPixelColor(pos + 1, strip.Color(r/2, g/2, b/2));
    strip.setPixelColor(pos + 2, strip.Color(r/4, g/4, b/4));

    strip.show();
    delay(wait);
    // If we wanted to be sneaky we could erase just the tail end
    // pixel, but it's much easier just to erase the whole thing
    // and draw a new one next time.
    for(j=-2; j<= 2; j++)
        strip.setPixelColor(pos+j, strip.Color(0,0,0));
    // Bounce off ends of strip
    pos += dir;
    if(pos < 0) {
      pos = 1;
```

```
      dir = -dir;
    } else if(pos >= strip.numPixels()) {
      pos = strip.numPixels() - 2;
      dir = -dir;
    }
  }
}

// Sine wave effect
#define PI 3.14159265
void wave(uint32_t c, int cycles, uint8_t wait) {
  float y;
  byte  r, g, b, r2, g2, b2;

  // Need to decompose color into its r, g, b elements
  g = (c >> 16) & 0x7f;
  r = (c >>  8) & 0x7f;
  b =  c        & 0x7f;

  for(int x=0; x<(strip.numPixels()*5); x++)
  {
    for(int i=0; i<strip.numPixels(); i++) {
      y = sin(PI * (float)cycles * (float)(x + i) / (float)strip.numPixels());
      if(y >= 0.0) {
        // Peaks of sine wave are white
        y  = 1.0 - y; // Translate Y to 0.0 (top) to 1.0 (center)
        r2 = 127 - (byte)((float)(127 - r) * y);
        g2 = 127 - (byte)((float)(127 - g) * y);
        b2 = 127 - (byte)((float)(127 - b) * y);
      } else {
        // Troughs of sine wave are black
        y += 1.0; // Translate Y to 0.0 (bottom) to 1.0 (center)
        r2 = (byte)((float)r * y);
        g2 = (byte)((float)g * y);
        b2 = (byte)((float)b * y);
      }
      strip.setPixelColor(i, r2, g2, b2);
    }
    strip.show();
    delay(wait);
  }
}
```

The Next Chapter

You've mastered breadboarding and soldering, and now it's time to kick things up a bit! Chapter 4, "Setting Up Wireless Connections," shows you how to create a quick wireless network. You'll then use this knowledge to create a simple doorbell for your house.

4

Setting Up Wireless Connections

This chapter explores the wireless networking tools that enable two or more Arduinos to talk together. Chief among these is the XBee, an Arduino-friendly wireless module capable of connecting a whole network of microcontrollers. In Figure 4.1, you can see one of my own projects, a LEGO robot controlled with Wii nunchucks connected to XBee-equipped Arduinos. You can learn how to build it in my book, *Make: Lego and Arduino Projects* (ISBN 978-1449321062). After you get up to speed on the XBee, you will tackle the third project, a wireless doorbell!

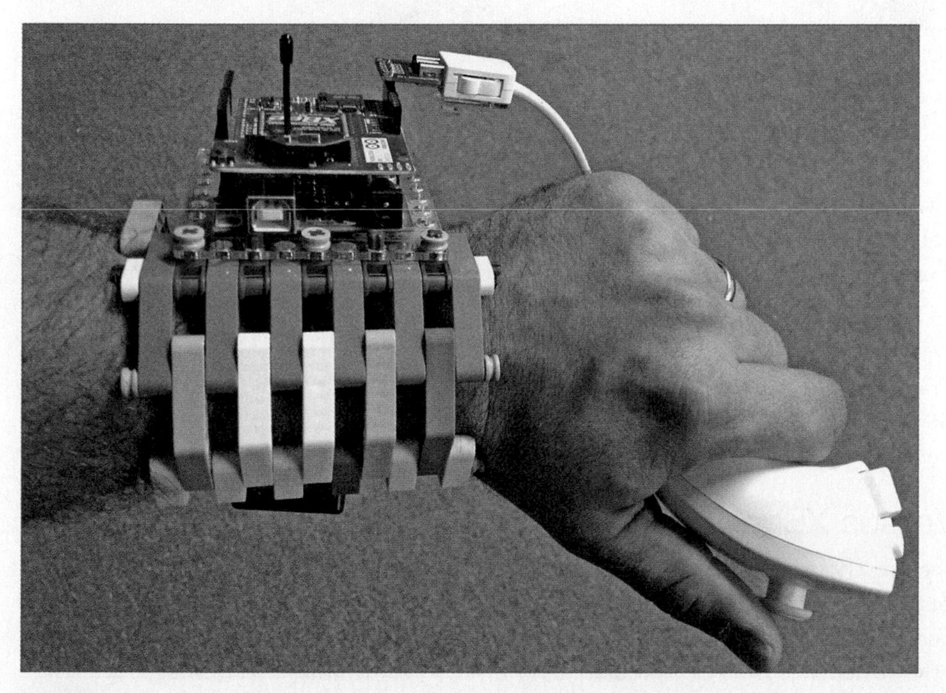

FIGURE 4.1 This XBee-equipped bracer enables you to control a robot wirelessly.

XBee Wireless Modules

XBee modules (see Figure 4.2) are based on ZigBee, which is an industry standard protocol that creates networks of multiple wireless nodes via serial data transmission, meaning only one bit (0 or 1) is sent at a time, making it slow but easy to configure. ZigBee is the default protocol used in home automation, so learning the platform's ins and outs could aid you in creating your own curtain-puller or light-switcher!

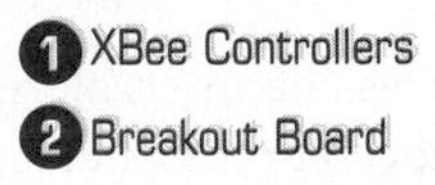

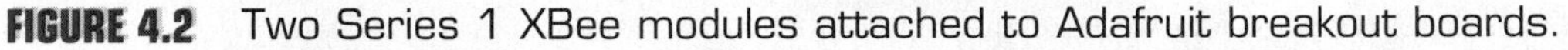
FIGURE 4.2 Two Series 1 XBee modules attached to Adafruit breakout boards.

XBee also happens to be the default communication method used by Arduino, enabling them to work together nicely. However, a wide assortment of XBee flavors are available, and you must sure to get the right one. Let's focus on just four of those XBee flavors in this chapter:

- XBee
- XBee Pro
- XBee Series 1
- XBee Series 2

XBee Versus XBee Pro

You first need to choose between XBee "regular" and "professional"—the distinction is purely about radio power. Ordinary XBees feature 1mW (one thousandth of a watt) power, whereas Pros are rated at 63mW, giving you a much greater range. What kind of range exactly? It depends on a complicated array of factors, including electromagnetic interference, antenna type, and physical obstructions.

That said, Digi International, the maker of XBee products, issues range estimates for the various models. The regular 1mW XBee is rated for 80 feet indoors and 300 feet outdoors, and the company claims the Pro model is good for 140 feet indoors and an impressive 4,000 feet—almost a mile—outdoors. Of course, for that last number, you would need the most ideal circumstances, like beaming from one hilltop to another. Any sort of obstruction will reduce the effective range of your radio.

If you don't need 4,000 feet, you might be better off skipping the Pro model because it costs more.

Series 1 Versus Series 2

The second consideration in choosing an XBee is what sort of networking you would like to configure. Digi International sells what it describes as Series 1 and Series 2 XBees.

- **Series 1**—Series 1 offers the simplest networking setup in that you don't have to set it up. Basically, every Series 1 module talks to every other Series 1 module within range—a configuration known as the mesh network. It's an easy way to get started playing around with wireless technology.

  If you want to direct data to a single module, you have to use software to set an identifier during both transmission and reception. This sounds intimidating, but it can be as simple as adding a single digit. Say you want to send data to Node 5; you can add a 5 to the beginning of your stream of data and the other nodes will ignore it.
- **Series 2**—Series 2 is more robust, offering—in addition to the settings of the Series 1—the ability to
  - Create more intricate networks with nodes being designated as "coordinators," able to issue commands.
  - Create "routers" that send and receive data.
  - Create end devices that may only receive.

  On the downside, having all these features means that you can't plug-and-play, because you must configure the modules before using them, unlike Series 1, which you can use right out of the box! More technically, the Series 2 use a different wireless protocol that makes them incompatible with Series 1 modules, so don't even try!

XBee Breakout Boards

XBee modules are easy to use, but they require a little love before they will fit into a typical Arduino project because their pin spacing is 2mm instead of Arduino-compatible 0.1". The solution is a small PCB called a *breakout board*, a way of creating a tiny circuit that can be plugged in to an Arduino.

The wimpiest of these is simply a PCB (printed circuit board) equipped with pins with the right spacing for breadboarding. However, more robust breakout boards, such as Adafruit's (P/N 126, previously shown in Figure 4.2), have a voltage regulator and status LEDs to keep your radio from getting fried.

Anatomy of the XBee

If you look at an XBee module, shown in Figure 4.3, it looks like a blue plate the size of a postage stamp, with a number of metal pins sticking out underneath. The top features an antenna. Adding it to a breakout board makes for more detail, so let's go through the XBee's various features.

FIGURE 4.3 The XBee and its breakout board breadboarded up. Note that the 5V and GND pins are already connected to the proper terminal buses.

1. **Pins**—You can see the tops of the XBee's pins. They control the board, bringing in power and sending and receiving data from the Arduino. The pins plug into headers on the breakout board. Note that these pins have the wrong spacing for breadboards.
2. **Antenna**—You have multiple antenna options depending on the XBee, but I think this wire antenna is the best for what it does, because it's tough and can take a modest amount of abuse without bending.
3. **Power LED**—This lights when the board powers up.

4. **Data LED**—This flashes to let you know that data is passing through the XBee.
5. **Power regulator**—These capacitors and the transistor manage the power going into the XBee. Unfortunately, frying a radio by using too much power is easy to do. The good news is that the regulator keeps the power flowing at just the right voltage.
6. **Breadboard pins**—Unlike the pins that connect the XBee to the breakout board, these pins are spaced correctly for a breadboard. Just as good, they are labeled so you can see which pin does what!

Competing Wireless Modules

It probably doesn't surprise you that the XBee isn't the only party in town. Here are a couple of cool alternatives that you can purchase for use in a project.

Freakduino Chibi

Created by Tokyo-based hacker Akiba (a.k.a. Chris Wang), the Chibi (see Figure 4.4) does away with the separate boards for the microcontroller and wireless module—Akiba has combined them into a single board. The Chibi is Arduino compatible and uses the same wireless band as the XBee. You can buy it at www.freaklabsstore.com.

FIGURE 4.4 Freaklabs' Freakduino Chibi is essentially an Arduino with built-in wireless capability.

JeeLabs JeeNode

A similar concept to the Chibi, the JeeNode consists of an ATmega328p, which is the same microchip that serves as the mind of the Arduino, along with a built-in wireless module. JeeNodes are very small and have fewer capabilities than the Chibi, but have many fans due to the JeeNodes' small form factor and their ease of use. You can purchase them at http://jeelabs.com/products/jeenode.

TIP

Just Use Series 1

There is so much more to learn about radios, and you might already be overwhelmed! I suggest just limiting yourself to the XBee, non-Pro, Series 1. It's a wonderfully simple way to add wireless to your projects without spending too much money or frustrating yourself by taking on too complicated a radio before you need to.

Project: Wireless LED Activation

Oooh, wireless radios! Working with them sounds kind of intimidating. It's actually not, and I'll prove it. Let's create a simple network (see Figure 4.5) that lets two Arduinos communicate. In this mini-project, you'll create two identical assemblies, each consisting of an Arduino and XBee, along with a button and a LED. When you press the button on one assembly, the LED on the other one lights up, and vice versa! You can see how this project will give you a nice start toward building a wireless doorbell, which is the main project for this chapter.

FIGURE 4.5 Control LEDs with XBee-equipped Arduinos.

PARTS LIST

You'll be making two assemblies, so you need two of everything!

- Arduinos (x2)
- XBees (x2)
- Breakout boards (x2)
- Pushbuttons (x2)
- Breadboards (x2)
- LEDs (x2)
- Jumpers

Follow these steps to assemble the XBee test platform:

1. **Solder the breakout boards**—Solder up your XBee breakout boards if you haven't already. Depending on your kit, this could mean simply soldering in some header pins. On other kits, however, you must solder in LEDs, capacitors, and so on.
2. **Connect the XBees to the breakout boards**—Attach the XBees to their respective breakout boards. This typically involves simply plugging in the XBees' pins to the appropriate holes in the breakout board. Just follow the directions that accompany your kit.
3. **Attach to breadboards**—Plug the breakout boards and XBees into the breadboards. You can see where to place it in Figure 4.6.
4. **Attach the pushbuttons, LEDs, and jumpers**—Attach these items as follows (also shown in Figure 4.6):
   A. GND on the XBee goes to GND on the breadboard. Connect the GND bus of the breadboard to the GND port of the Arduino.
   B. +5V on the XBee goes to 5V on the Arduino.
   C. TX on the XBee goes to RX on the Arduino.
   D. RX on the XBee goes to TX on the Arduino.
   E. Connect a button to pin 8 on the Arduino; the other end connects to the GND bus.

You should end up with two identical units, and if you upload the Arduino code to both of them, they should work identically. Even cooler, the way the networks are set up, you could actually create three or more of these assemblies and they'll all work the way you would expect. Press the button on one, and the LEDs on all the others will light up! It's not super practical, to be sure, but it shows how easily you can set up an XBee network.

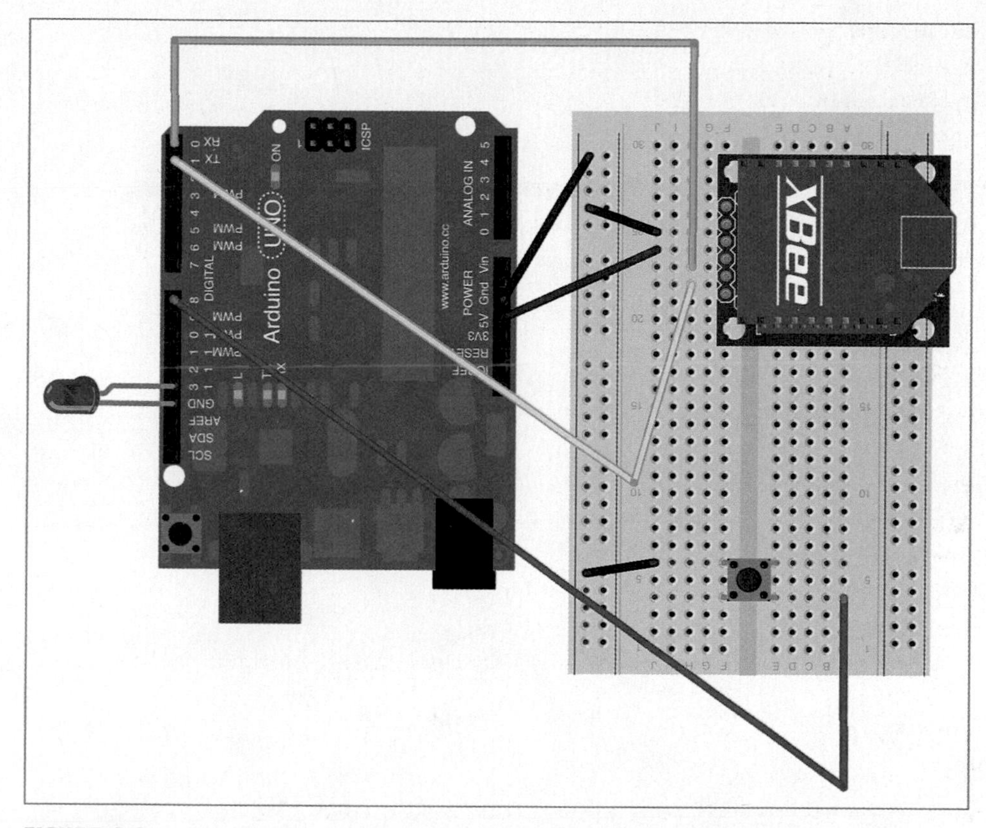

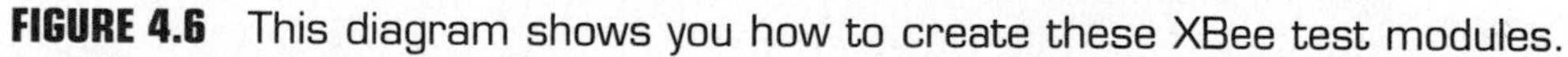

FIGURE 4.6 This diagram shows you how to create these XBee test modules.

Wireless LED Code

Upload the following code to both Arduinos. Remember, both modules are identical, down to the software. If you can't remember how to upload sketches to your Arduino, Chapter 5, "Programming Arduino," explains how.

> **NOTE**
>
> **Code Available for Download**
>
> You don't have to enter all of this code by hand. Simply go to https://github.com/n1/Arduino-For-Beginners to download the free code.

```
#include <Wire.h>

const int buttonPin = 8;
const int ledPin = 13;
int buttonState = 0;

void setup()
{
  Serial.begin(9600);
  pinMode(ledPin, OUTPUT);
 pinMode(buttonPin, INPUT_PULLUP);
}

void process_incoming_command(char cmd)
{
  int speed = 0;
  switch (cmd)
  {
  case '1':
  case 1:
    digitalWrite(ledPin, LOW);
    break;
  case '0':
  case 0:
    digitalWrite(ledPin, HIGH);
    break;

  }
}


void loop() {
  if (Serial.available() >= 2)
  {
    char start = Serial.read();
    if (start != '*')
    {
      return;
    }

    char cmd = Serial.read();
    process_incoming_command(cmd);
```

```
  }

  buttonState = digitalRead(buttonPin);
  if (buttonState == HIGH) {
    Serial.write('*');
    Serial.write(1);
  }
  else {
    Serial.write('*');
    Serial.write(0);
  }
  delay(50); //limit how fast we update
}
```

Project: Bluetooth Doorbell

Now you can take what you learned about XBees and apply it to a slightly more robust project: a wireless doorbell. Figure 4.7 shows the doorbell button, and Figure 4.8 shows the buzzer unit that is tucked away on a shelf inside.

FIGURE 4.7 The doorbell awaits visitors!

FIGURE 4.8 The buzzer unit sits discreetly on a shelf.

Sure, you might say, they make these already! You can buy a wireless doorbell in any hardware store. However, this one you make yourself! Even better, as you get more confident with Arduino, you can modify it to make it uniquely yours. For instance, what if your Arduino triggers a music player instead of a buzzer to let you know that someone has pressed the button?

PARTS LIST

Just as in the mini-project earlier in the chapter, you'll be using two Arduinos, linked together. However, in this project, one Arduino waits for a button press, while the other one sets off a buzzer when it detects that the button has been pressed.

- 2 Arduino Unos
- 2 XBee wireless modules (Adafruit P/N 128)
- 2 Adafruit XBee breakout boards (Adafruit P/N 126)
- 2 mini breadboards (these are really small breadboards the sign of a postage stamp, Adafruit P/N 65)
- Button (SparkFun P/N COM-10443)
- A 330-ohm resistor
- Buzzer (Jameco P/N 1956776)
- Jumpers
- 9v battery clip (Jameco P/N 105794)
- 9v connector with barrel plug (Adafruit P/N 80)
- 1/4-inch MDF for enclosure backing and sides
- 5mm acrylic for enclosure front
- 1-inch #4-40 bolts
- Hot glue gun

The Button

The button you use in the button unit, shown in Figure 4.9, is kind of intriguing because it has six connectors: two sets of positive and negative terminals that close when the button is pressed—so you could have two circuits, both of which trip when the button is activated. The last two leads—the white lugs in the photo—are for powering the LED. Be sure to attach a resistor on the power lead so you don't fry your LED inadvertently. I use a 330-ohm resistor in this project.

FIGURE 4.9 The button you use in the project has six connectors.

Instructions for Wiring Up the Doorbell

The project consists of two Arduinos equipped with XBee modules and breakout boards. One Arduino has a button, and the other has a buzzer to sound out to let you know someone is at your door. Let's get started!

Button Unit

Let's begin with the button unit (see Figure 4.10), which consists of the following components:

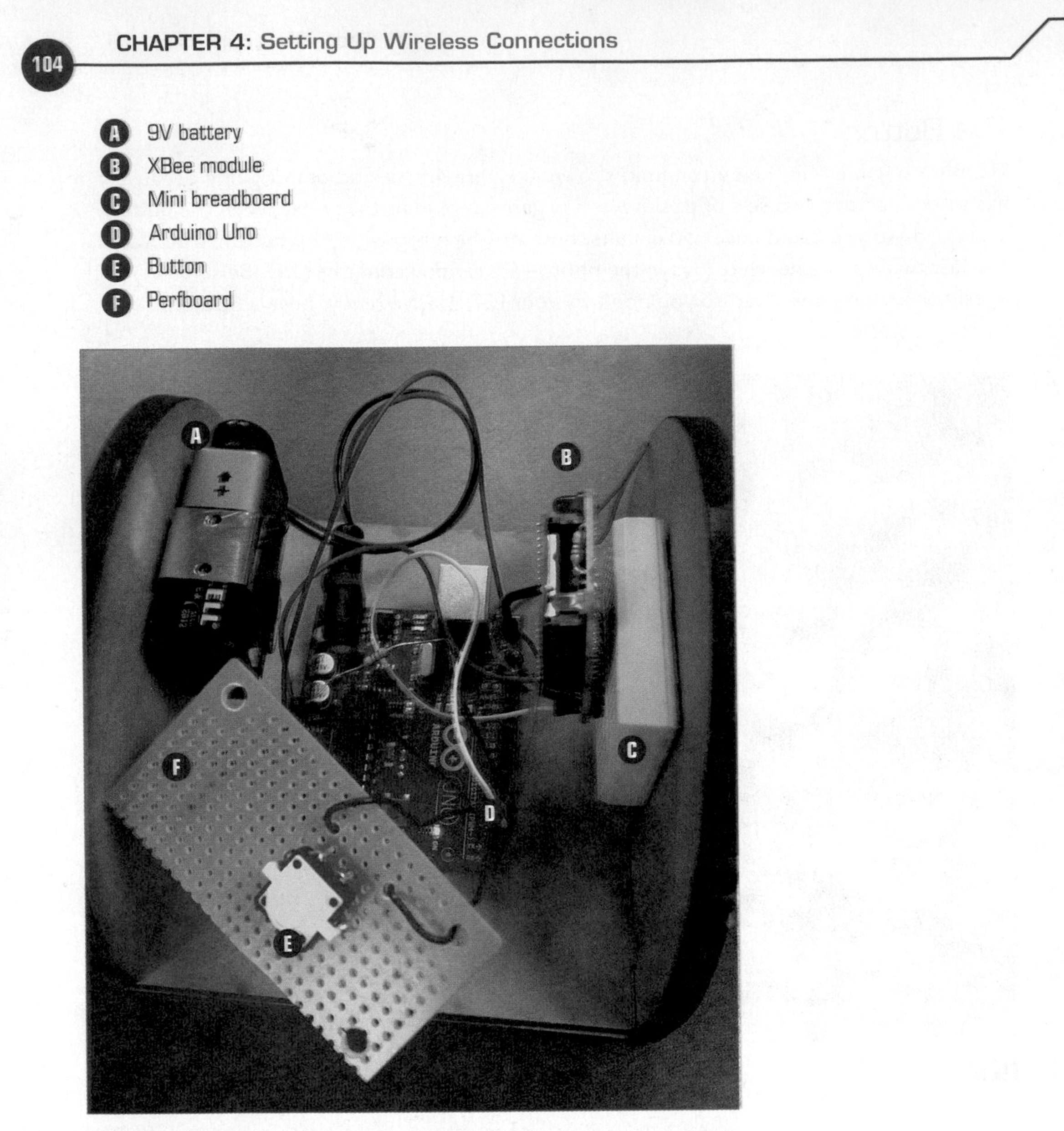

FIGURE 4.10 The button unit before the acrylic is added.

Now, assemble these parts together as shown in Figure 4.11, and you can follow along with these steps:

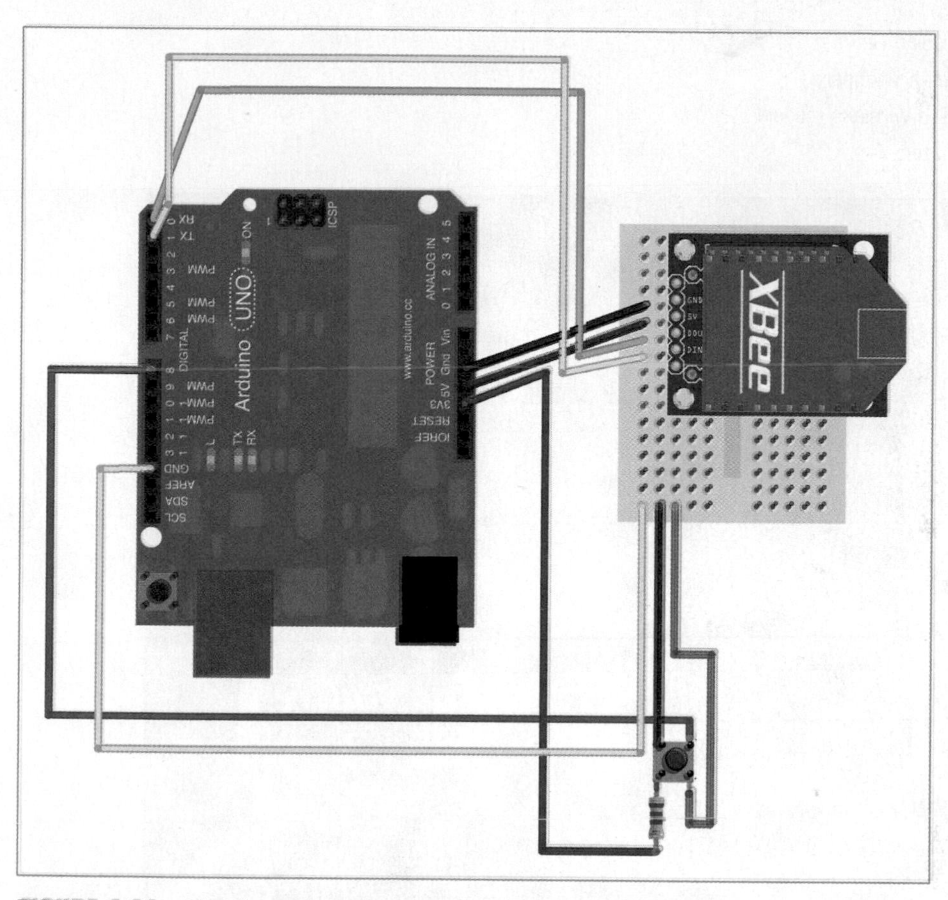

FIGURE 4.11 The button unit consists primarily of a button, an Arduino, and the wireless module.

1. Plug in the XBee and its breakout board to a mini breadboard.
2. Plug the XBee's 5V to the 5V on the Arduino, its TX into RX, its RX into TX, and its GND pin to any free GND on the Arduino.
3. Connect one of the button's leads to pin 8 and the other to GND. (I use the breadboard to accommodate the GND leads coming from the button.)
4. Solder a 330-ohm resistor and a jumper to the button's LED's power terminal, and connect the other end to the 3V3 port of the Arduino. The other terminal of the LED goes to GND.

Buzzer Unit

Next, connect the components that make up the buzzer unit, seen in Figure 4.12. These consist of the following:

- **A** Arduino Uno
- **B** Mini breadboard
- **C** XBee wireless module
- **D** Buzzer

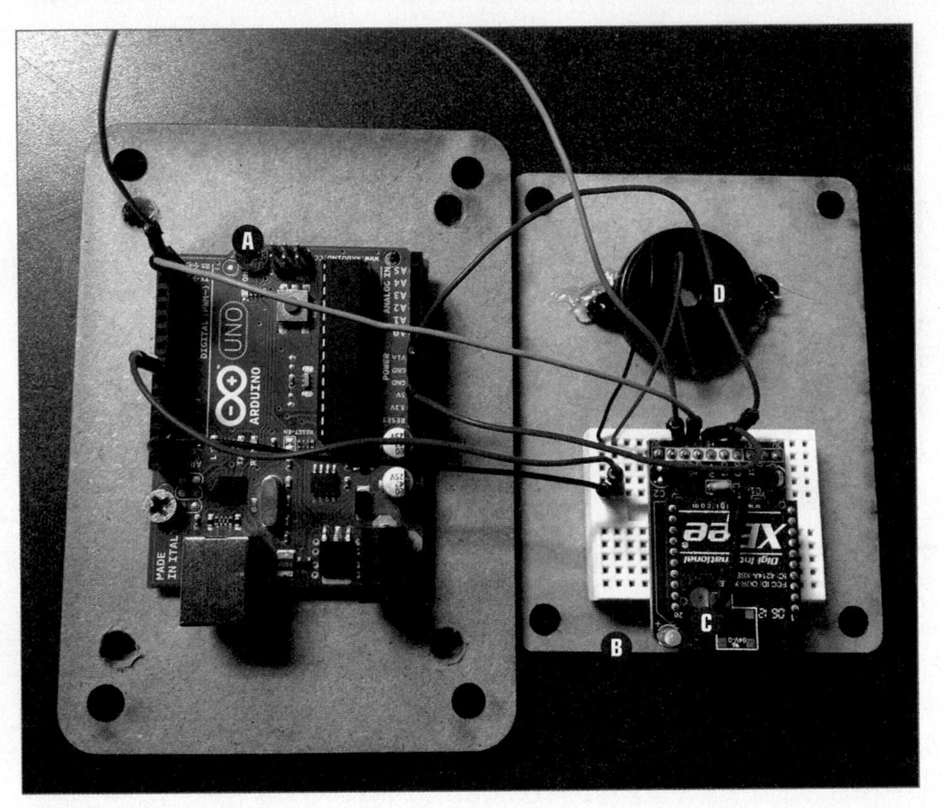

FIGURE 4.12 The buzzer unit waiting to be closed up. The outer holes are for wall mounting.

Next, use Figure 4.13 as a guide for connecting the various parts:

1. Plug in the XBee and its breakout board into a mini breadboard.

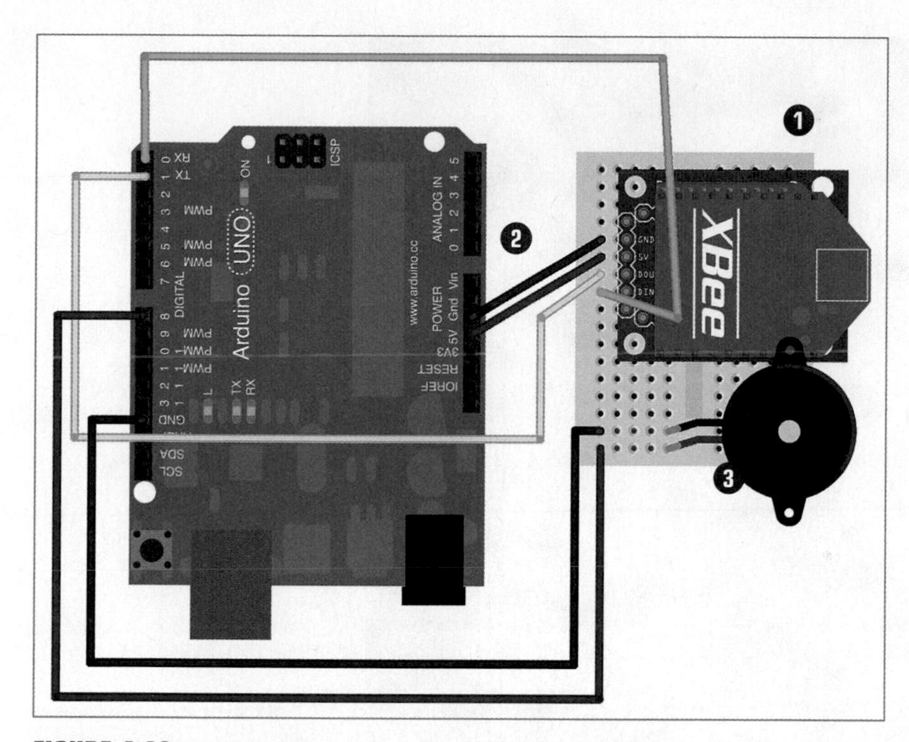

FIGURE 4.13 The buzzer unit consists of an Arduino, XBee, and buzzer.

2. Plug in the XBee's 5V to the 5V on the Arduino, its TX into RX, its RX into TX, and its GND pin to any free GND on the Arduino.
3. Connect the buzzer's leads to the breadboard as well, as shown in Figure 4.13. You can connect them directly to the Arduino if you want—if you go this route, connect the red wire to pin 8 and the black wire to any free GND.
4. To power the buzzer unit, use an Arduino-compatible wall wart or a 9V battery pack.

Building the Doorbell Enclosures

You next need to build the two enclosures for this project. The outside enclosure (see Figure 4.14) is designed to resist the elements—I hesitate to call it "weatherproof"—whereas the inside enclosure is designed to look good.

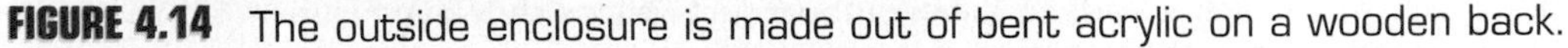

FIGURE 4.14 The outside enclosure is made out of bent acrylic on a wooden back.

Button Unit Enclosure

The button unit is the module that is on the outside of the door—press the button to make the buzzer buzz! To make an enclosure, all you need is a box with a hole for the button, but I'll show you how you can make one of your own. The one I made consists of a sheet of acrylic that I bent by heating it up, and then laying the flexible acrylic over a metal pipe to form a half-circle. I added the acrylic to a wooden back (refer to Figure 4.11) to finish the enclosure. Here are the steps:

1. Laser-cut the top, bottom, and back out of quarter-inch medium-density fiberboard (MDF). If you don't have access to a laser cutter, you can create a box out of pieces of wood, repurpose another container as an enclosure, or buy a commercial project enclosure.
2. Laser-cut the front from 5mm acrylic. (If you want the design files I used to output the wooden backing as well as the acrylic front, you can find them at https://github.com/n1/Arduino-For-Beginners.)

3. Glue the top and bottom wood pieces to the back wood piece. You might want to paint the wood!
4. Attach the completed electronics as shown earlier in Figures 4.10 and 4.12. Use the #4-40 bolts for the Arduinos and hot glue for the buzzer, battery pack, and mini breadboards.

   If you aren't using a laser cutter, you'll need to drill mounting holes in the acrylic. You might want to mock it up using a sheet of paper first.
5. Bend the acrylic front plate as described in the next section, "Bending Acrylic."
6. Attach the acrylic plate to the front so that the button can be pressed through the hole in the plastic.
7. Install the unit outside your door of choice, and eagerly await your first visitor!

Bending Acrylic

For the outside button unit enclosure, you heat-bend acrylic (see Figure 4.15) to form a casing. This task is easy to learn because you don't really need anything unusual or uncommon.

FIGURE 4.15 Bending acrylic is easy and gives a nice effect!

Acrylic (also known as Plexiglas) is also easy to heat and re-form. After it gets to the right temperature—not too hot or cool—the acrylic starts to bow and flex. When it gets a little hotter, it softens. That's when you bend it how you want it, and let it cool into an awesome new shape!

You need three things to get started:

- **The acrylic to be bent**—I suggest 1/8 inch, though you might have luck with the thicker stuff.
- **A form**—This is the surface over which the hot acrylic will cool and harden. You want this close to the actual curve you want the plastic to hold. The easiest form of all is the edge of a table. I used a rounded form—a pipe—to form the acrylic face seen in Figure 4.7. If you go this route, you'll need to find a form that matches the curve of the shape you're looking for.
- **A source of heat**—Heat guns (see Figure 4.16) and propane torches are common tools, though you can purchase commercial acrylic-heating strips (TAP Plastics has one for $80, P/N 169). Finally, you could heat up the plastic in an oven. This last technique is not for the faint of heart and you should definitely monitor the plastic closely so it doesn't bubble or scorch.

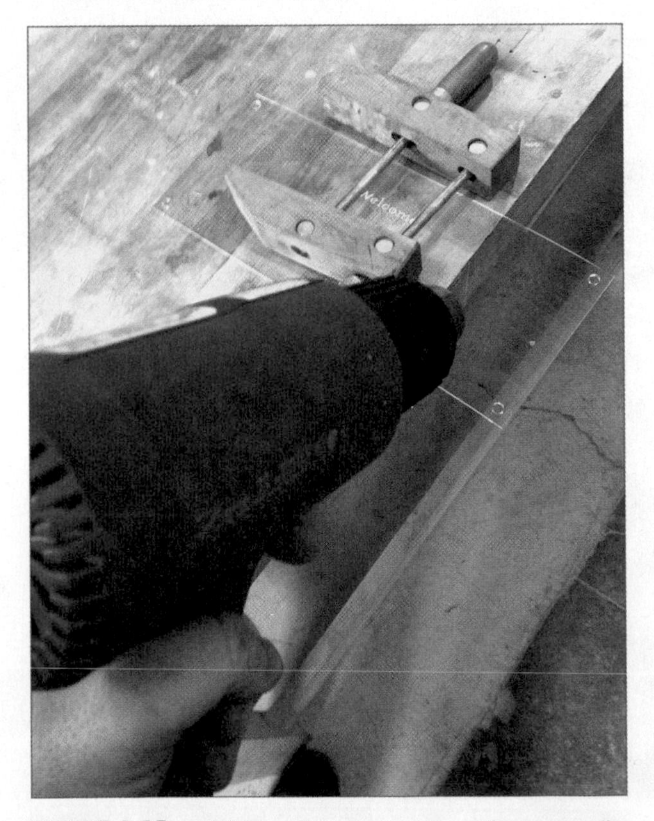

FIGURE 4.16 Using a heat gun to soften acrylic.

Although you could conceivably use any heat-resistant surface to form your acrylic—or even build your own out of pieces of wood—in some respects, using the edge of the table is an easy choice because it bends the plastic perfectly, using gravity and the table's surface to make a fairly perfect 90-degree bend. To bend plastic using the "edge of the table" technique, follow these steps:

1. As shown in Figure 4.17, position the acrylic so the edge of the table is right where you want the plastic to bend. You'll definitely want to weigh it down so it doesn't move.

FIGURE 4.17 As the acrylic heats up, it starts to bend.

   When it gets hot enough, gravity starts pulling the soft acrylic down, as shown in Figure 4.17.

2. Position the acrylic how you want it to look—and work quickly because after it cools, it becomes just as brittle as it was before. Don't try to re-bend it without applying more heat!

Buzzer Unit Enclosure

The buzzer unit doesn't use plastic, because who wants plastic in their home? Instead, you can use a simple arrangement of wooden panels separated by bolts. I laser-cut two pieces of wood, one bigger than the other. (I ended up hand-drilling four additional holes, as shown in Figure 4.18, after changing my mind on how to proceed.)

FIGURE 4.18 I used laser-cut wood for the buzzer unit's enclosure.

To connect the two pieces I used brass bolts, #10-24 and 2.5" long, with brass washers and nuts. This enclosure is considerably easier to do than the other enclosure and it looks great!

Wireless Doorbell Code

Upload the following code to your Arduinos. If you're having difficulty figuring out how to upload your sketches, see Chapter 5 to learn how. As before, you can download the code from https://github.com/n1/Arduino-For-Beginners.

Button Unit Code

The Button Unit sketch consists of a loop that waits for the button to be pressed, then transmits a wireless alert.

NOTE

Code Available for Download

You don't have to enter all of this code by hand. Simply go to https://github.com/n1/Arduino-For-Beginners to download the free code.

```
#include <Wire.h>

const int buttonPin = 8;
int buttonState = 0;

void setup()
{
  Serial.begin(9600);
 pinMode(buttonPin, INPUT_PULLUP);
}

void loop() {
  if (Serial.available() >= 2)
  {
    char start = Serial.read();
    if (start != '*')
    {
      return;
    }

    char cmd = Serial.read();
  }

  buttonState = digitalRead(buttonPin);
  if (buttonState == HIGH) {
    Serial.write('*');
    Serial.write(1);
  }
  else {
```

```
    Serial.write('*');
    Serial.write(0);
  }
  delay(50); //limit how fast we update
}
```

Buzzer Unit Code

The Buzzer Unit code is similarly plain. The loop monitors serial traffic, then sounds the buzzer when it detects the command from the Button Unit.

> **NOTE**
>
> **Code Available for Download**
>
> You don't have to enter all of this code by hand. Simply go to https://github.com/n1/Arduino-For-Beginners to download the free code.

```
#include <Wire.h>

const int buzzerPin = 13;

void setup()
{
  Serial.begin(9600);
  pinMode(buzzerPin, OUTPUT);
}

void process_incoming_command(char cmd)
{
  int speed = 0;
  switch (cmd)
  {
case 1:
    digitalWrite(buzzerPin, LOW);
    break;
case 0:
    digitalWrite(buzzerPin, HIGH);
    break;

  }
}
```

```
void loop() {
  if (Serial.available() >= 2)
  {
    char start = Serial.read();
    if (start != '*')
    {
      return;
    }

    char cmd = Serial.read();
    process_incoming_command(cmd);
  }

  delay(50); //limit how fast we update
}
```

The Next Chapter

So far we've been talking the hardware angle, but now it's time to switch things up! You get to delve into Arduino code in Chapter 5 and learn a bunch of programming techniques as well as the specific formatting you'll need to successfully write your very own Arduino program.

5

Programming Arduino

So far you've gotten a taste of programming, if only by uploading code to your Arduino board, pictured in Figure 5.1. In this chapter, you explore main areas of programming and learn techniques to help you master Arduino programming on your own.

FIGURE 5.1 The Arduino programming environment enables you to control the Arduino.

NOTE

Code Available for Download

You don't have to enter all of this code by hand. Simply go to https://github.com/n1/Arduino-For-Beginners to download the free code.

The Arduino Development Environment

The software used to program Arduinos (confusingly called Arduino, too!) is called an integrated development environment or IDE. Most of the time when a person creates a program, it is within the framework of this environment. In layperson's terms, you need a program running—the Arduino IDE—to successfully send code to your board. Let's go over this environment and learn about its details.

Programming Window

Broadly speaking, the IDE consists of a programming window and a set of menus. Figure 5.2 shows the Arduino IDE environment. Follow along with the callouts to see what each option does.

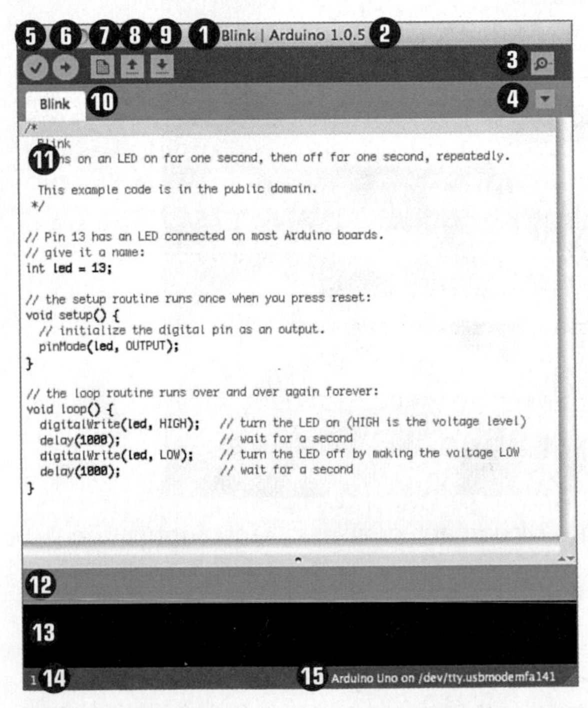

FIGURE 5.2 The Arduino environment's interface gives users a lot of options.

1. **Filename**—The filename of the sketch (the Arduino term for *program*) you're looking at—in this case, "Blink." New filenames default to the month and day.
2. **Version number**—Confused as to which version of Arduino you're running? Look at the number to see what version number is loaded onto your computer. As always, be sure to keep your Arduino environment up to date, because previous versions might have bugs that can hinder you.
3. **Serial monitor**—Activates the serial monitor feature of the environment. I'll explain all about the serial monitor and how it could benefit you later in this chapter.
4. **Tab selector**—When you have multiple sketches open, you can use this button to navigate to the sketch you want to use.
5. **Verify**—Click here if you want to check the syntax of your code without either saving or uploading it to your Arduino.
6. **Upload**—Your best friend! This button verifies your code, compiles it into the programming language the Arduino understands, and sends it to the Arduino.
7. **New**—Create a new sketch. Clicking this button won't affect any other sketches you might have open. The new filename defaults to the current month and day.
8. **Open**—Opens up your "Arduino sketchbook," which is your default folder for sketches. From there, you can open files as you normally would.
9. **Save**—Save the sketch! Note that Arduino sketches must be saved within a folder, so if no folder exists for a new file, the environment will create one.
10. **Tab**—Each window has its own filename, which duplicates from the top of the window. Additionally, sometimes sketches consist of more than one tab. Load the example sketch `toneMelody` (see below for how to find example sketches) and you will see how this works. The second example file, `pitches.h`, is a source file that the main sketch needs to operate.
11. **Workspace**—This part of the window shows the sketch. If you haven't loaded a sketch, the workspace area is blank. This is your main workspace when creating a sketch.
12. **Status bar**—This blue bar displays status messages from the environment, usually regarding uploads. It will also display a progress indicator letting you know where you are in the upload process.
13. **Status window**—This black pane displays error messages incurred during upload. The Arduino software verifies all code being sent and won't upload if there's an error. If a problem is found, the error message attempts to display what's wrong, though it's often cryptic! As you can see, no error message appears in Figure 5.2.
14. **Line number**—Each line of code is numbered, and this indicator helps you navigate by constantly indicating which line the cursor is on.
15. **Board and port number**—A bunch of different Arduino boards exist, notwithstanding the fact that I exclusively use the Uno in this book. Select the board you're uploading to unless you like getting cool error messages! The second piece of information here is the port designator. This is an arcane technical subject, and you don't have to worry about it because the software will suggest one for you.

Menus

Now let's go over the various menu options, some of which are fairly typical (Open File), whereas others duplicate some of the options from the programming window. Still, they contain a lot of important stuff! Note that you will encounter minor differences in the menus based on whether you're running Mac OS X, Linux, or Windows. Just explore the menus and find out what each option does.

Figure 5.3 shows you what you get with the Mac OS X version of the File menu. If you use Linux or Windows, don't worry; the menus are pretty much the same. Let's go over what you get.

New	⌘N
Open...	⌘O
Sketchbook	▶
Examples	▶
Close	⌘W
Save	⌘S
Save As...	⇧⌘S
Upload	⌘U
Upload Using Programmer	⇧⌘U
Page Setup	⇧⌘P
Print	⌘P

FIGURE 5.3 The File menu helps manage your Arduino sketches.

File Menu

The typical options found in File menus in most any application are **New**, **Open**, **Close**, **Save**, and **Save As**, and they work how you would expect. Following are explanations for the other commands found in the File menu:

- **Sketchbook** opens a popup window of your sketchbook, giving you the option to launch one of your existing sketches—basically, the same option as Open icon in the programming window.
- **Examples** introduces you to example sketches, code that you're allowed to freely use and adapt to make your own programs. I discuss example text later in the chapter in the section, "Learning from Example Code."
- **Upload** sends the sketch in the programming window to the Arduino board, assuming they're connected!
- **Upload Using Programmer** is an advanced option. Don't mess around with this, because you could potentially wipe the Arduino's firmware!

Edit Menu

Clicking the Edit menu shows the list of options in Figure 5.4. As with the File menu, you will see a number of typical options that hardly require an explanation, such as **Cut**, **Copy**, **Paste**, **Select All**, as well as the various **Find** commands.

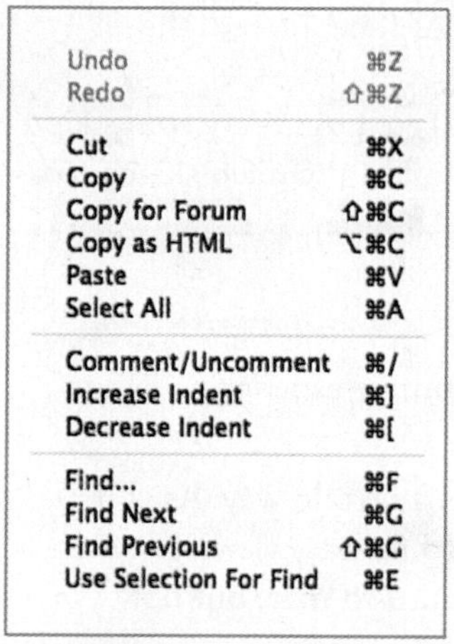

Undo ⌘Z
Redo ⇧⌘Z
Cut ⌘X
Copy ⌘C
Copy for Forum ⇧⌘C
Copy as HTML ⌥⌘C
Paste ⌘V
Select All ⌘A
Comment/Uncomment ⌘/
Increase Indent ⌘]
Decrease Indent ⌘[
Find... ⌘F
Find Next ⌘G
Find Previous ⇧⌘G
Use Selection For Find ⌘E

FIGURE 5.4 The Edit menu gives you the usual Copy and Paste, along with some fun extras!

The Edit menu has a number of intriguing options:

- **Copy for Forum** enables you to copy Arduino sketches with formatting. In the case of Copy for Forum, the text is formatted to look good when pasted into Arduino's forums.
- **Copy as HTML** copies a sketch and adds the appropriate tags so the sketch looks great on your web page. It's pretty slick!
- **Comment/Uncomment** changes a line or block of text so that it's commented, which means that Arduino doesn't recognize the text as actual code. I talk more about commenting later in this chapter, in the section "The Blink Sketch."
- **Increase Indent** and **Decrease Indent**, well, change the indentation of code. If you look at an Arduino sketch, you can see it uses indents to help organize the code for easy viewing.

Sketch Menu

The Sketch menu (see Figure 5.5) offers a couple of options for managing sketches—remember, sketches are what you call Arduino programs. **Verify/Compile** verifies the code and compiles it. Because the environment won't compile bad code, this is a great way to ensure that your sketch is very likely to work! I say "very likely" because even though the code compiles, it doesn't mean that it works the way you want it to.

Verify / Compile ⌘R
Show Sketch Folder ⌘K
Add File...
Import Library... ▶

FIGURE 5.5 The Sketch menu offers options for connecting libraries and support files to your sketch.

- **Show Sketch Folder** opens the folder for the current sketch. As I mentioned, the environment wants every sketch to have its own folder.
- **Import Library** adds the reference code for a library to your sketch. A library works kind of like a source file in that it's a file with additional code left off of the main sketch for simplicity's sake. We'll explore libraries later, in the section, "Libraries."

Tools Menu

Tools are just that, utilities for helping you manage your programming experience. Figure 5.6 shows the available options on the Tools menu.

- **Auto Format** arranges each line of the sketch so that it looks a certain way. As you open the example sketches, you can see that they're formatted for easy viewing. When you create your own programs, the lines you type often will auto-format, but not always!

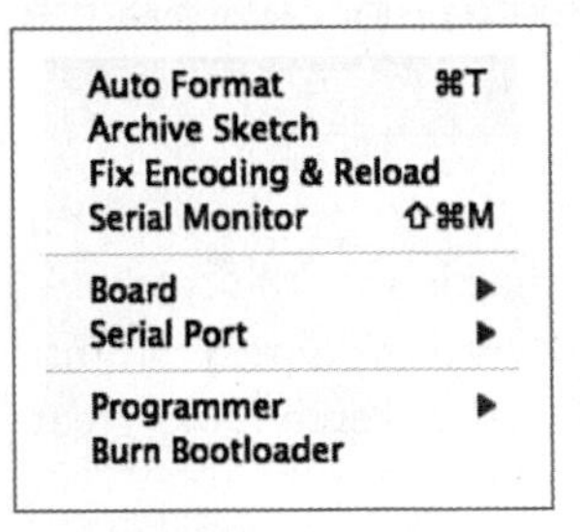

FIGURE 5.6 The Tools menu has a bunch of options for making your sketches work better.

- **Archive Sketch** creates a compressed .ZIP file with your sketch in it.
- **Fix Encoding & Reload** fixes typographical problems with text—for instance, in some cases, a smart quote or special accent character displays as a code instead of a single character. This function helps fix those translation problems.
- **Serial Monitor**, like the icon mentioned earlier in this chapter, under "Programming Window," activates the environment's serial monitor tool. This tool is extremely helpful and we delve into it in the section called "Debugging with the Serial Monitor."

NOTE

Programmer Tool Not Covered Here

The Programmer tool is an advanced topic that falls outside the bounds of this book. However, you can learn more about this advanced topic in *Sams Teach Yourself Arduino Programming in 24 Hours*, by Rich Blum, slated to publish in spring 2014.

- **Board** lets you choose which Arduino board you use. Throughout this book only Arduino Unos are used, but there are many varieties of Arduino, and you'll have to select the right one to successfully upload your sketch. If you forget and select the wrong board, the upload will fail with an error message.
- **Serial Port** gives you the option of choosing which serial port you would like to use to upload your sketch. This is fairly cryptic business, and chances are you won't need to tinker with this setting unless you are an advanced user.
- **Burn Bootloader** is the tool that formats the Arduino's microcontroller chip. Again, this is an advanced topic that we don't have space for.

Help Menu

Arduino's Help menu (see Figure 5.7) consists of a series of documents stored on the computer with the Arduino software. They're the typical top-level help documents, but they're well worth a look. In the section "Debugging with the Serial Monitor," you learn various ways to troubleshoot your sketches, but reading these files is a start.

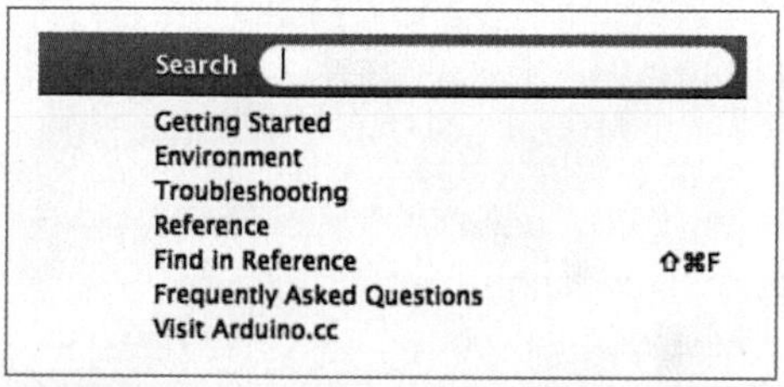

Search

Getting Started
Environment
Troubleshooting
Reference
Find in Reference ⇧⌘F
Frequently Asked Questions
Visit Arduino.cc

FIGURE 5.7 Only a handful of documents are in the Help menu, but they're all extremely useful!

If you want to learn more about the Arduino IDE, check out Arduino's page on the subject at http://arduino.cc/en/Guide/Environment. Now that you've been introduced to the menu system, let's look at a sketch!

UPLOADING SKETCHES TO YOUR ARDUINO

What if you don't need all this info, and just want to learn how to upload your sketch (program) to the Arduino? This sidebar is for you. Here's how it works:

1. Make sure you have the latest version of the Arduino software.
2. Connect an Arduino to your computer via a USB cable. If you need a cable, Adafruit has a nice short one, P/N 900. Sometimes you get a popup window announcing that it has detected the Arduino; you can dismiss this.
3. Check to make sure you have selected the correct Arduino from the Tools > Board menu. You can read more about this menu from the Arduino Development Environment section this chapter.
4. Pull up your sketch in the Arduino software. Click Upload to send it to the board. You'll see the LEDs marked TX and RX flashing crazily as the code uploads. If no error messages result, the code should be on the Arduino and will run automatically.
5. If there is an error in your sketch, it won't upload. If you want to test out your connection with trusted code, send an example sketch from File > Examples. If it doesn't work, chances are you need to check your settings and connections.

The Blink Sketch

One of the fun traditions of computer programming is that usually, the first program you create prints the words "Hello, world!" This is the electronic equivalent of making an LED flash—and that's what the Blink sketch does.

Let's delve into the code. First, however, make sure you have the latest version of the Arduino software. Simply launch the application and look at the top of the programming window (callout 2 on Figure 5.2) to see the version number. Next, go to the Arduino website at arduino.cc and click on Download. It'll tell you the current release number there.

After you're sure you have the latest version, launching it is easy: Just choose **File** > **Examples** > **01.Basics** > **Blink**. The code appears in a window and you can edit it all you want, and then save it as a new filename. (You can't save over the Blink example; it's read-only.)

So, let's look at the code step by step:

NOTE

You Already Have This Code

I have not included this code sample for download from the book's website because the code for the Blink sketch is included with the Arduino IDE.

```
/*
  Blink
  Turns on an LED on for one second, then off for one second, repeatedly.

  This example code is in the public domain.
 */
```

As you can see, the first part of the Blink sketch is some information with a forward slash and asterisk and ends with them reversed. This is how you mark up text so that the Arduino ignores it. Generally, programmers insert this kind of text into their code to remind them what the code was designed to do, why they wrote it. Sometimes, it's also included as a teaching aid to other programmers.

Wouldn't it be funny if the Arduino mistook a note you left in the code as an instruction? Actually, no it wouldn't! It would be a headache for you to figure out what you did wrong. Therefore, if you want to leave a block of information in your code, use these tags.

```
// Pin 13 has an LED connected on most Arduino boards.
// give it a name:
int led = 13;
```

This next block of code has three interesting things going on. First, look at the double-slash preceding two of the lines. This is another way of writing a comment. In this case, the entire line following the double-slash is ignored by the Arduino, letting you leave important notes for yourself or other people, such as what Pin 13 is for!

The second thing to notice about the block of code is the third line. It's declaring a variable. `INT` refers to an integer, another way of saying a number. For reasons too complicated to get into here, the number must fall within a range of –32,768 to 32,767. If it doesn't, you have to use different syntax to declare your variable.

The word `led` is the actual variable name. As you might recall, the program flashes an LED on Pin 13. But for the Arduino to know this, you have to declare that Pin 13 is the one to flash. Declaring this takes two steps. The first step, shown in this block, is to create a variable called `led` with a value of 13. If you want to learn more about declaring variables, the Arduino website has a nice tutorial at http://arduino.cc/en/Reference/VariableDeclaration.

The third item of interest is the semicolon. Typically, every line of code ends in a semicolon, and that's how the Arduino knows to move on to the next line. If you miss one, an error message results and you won't be able to upload your code.

```
// the setup routine runs once when you press reset:
void setup() {
  // initialize the digital pin as an output.
  pinMode(led, OUTPUT);
}
```

This next block begins with an intriguing command: `void setup`. This is actually two commands in one. `Void` is a keyword that tells the Arduino that the information in this block is self-contained and doesn't send any information elsewhere in the larger program—this is not really an issue in a small program like Blink.

As the comment references, the `setup` block runs only once when the Arduino powers up or if the reset button is pressed, making it ideal for setting variables and other one-time-only tasks.

Next is a curly brace, which is a bracket that looks like `{`, as well as its mate, `}`, which appears at the end of the block of code. These are parts of the `void setup` command. Any functions within the braces are triggered at the same time as the main command.

In the second half of the variable declaration, the function `pinMode` sets an individual Arduino pin to either send data (output) or accept data (input). In this case, `pinMode` is telling the Arduino that pin `led` (which was previously set to Pin 13) is set to `output`. This means that the Arduino will trigger Pin 13 as instructed by the program, without attempting to get a reading from a hypothetical sensor plugged into the pin.

The final element of this block is the closing curly brace. Every time you use curly braces, they must have both an opening and a closing brace. If you don't have your braces "balanced" as this is known, your program will fail.

```
// the loop routine runs over and over again forever:
void loop() {
  digitalWrite(led, HIGH);   // turn the LED on (HIGH is the voltage level)
  delay(1000);               // wait for a second
  digitalWrite(led, LOW);    // turn the LED off by making the voltage LOW
  delay(1000);               // wait for a second
}
```

The final piece of the Blink code is the loop. Unlike `setup`, which runs only once, a loop runs repeatedly until the Arduino shuts down. This is typically where the functional part of the program resides.

The loop contains the command to activate the LED and deactivate it. As you can see from the comments, the `digitalWrite` function controls this task. Reading data from a pin or writing to it is controlled by this command and its cousins, `digitalRead`, `analogRead`, and `analogWrite`. As mentioned, any pin can be set to input and output. Additionally, a pin must be designated as either digital or analog. Chapter 6, "Sensing the World," explores what this means precisely, but suffice it to say that in this code, you control an LED with a `digitalWrite` command.

The keywords `HIGH` and `LOW` in the Arduino world refer to delivering voltage to the pins. Obviously, `HIGH` is on and `LOW` is off. This loop turns on the LED on Pin 13, and then waits one second—the `delay(1000);` command tells the Arduino to wait for 1,000 milliseconds—and then turns off the LED for another second, and then starts the loop again.

So you see, even in arguably the simplest program out there, you can still learn a lot of interesting stuff!

Learning from Example Code

You can find tons of example sketches by choosing **File** > **Examples** in the Arduino software's menu system (see Figure 5.8). Many of the examples cover basic stuff, but many of them serve as useful tutorials on advanced topics. This highlights the value of example code—not so much as a readymade solution, but more of a reference for learning the syntax to adapt for your own uses. For instance, choose **File** > **Examples** > **Digital** > **Button**. This sketch shows you how to light up an LED by pressing a button.

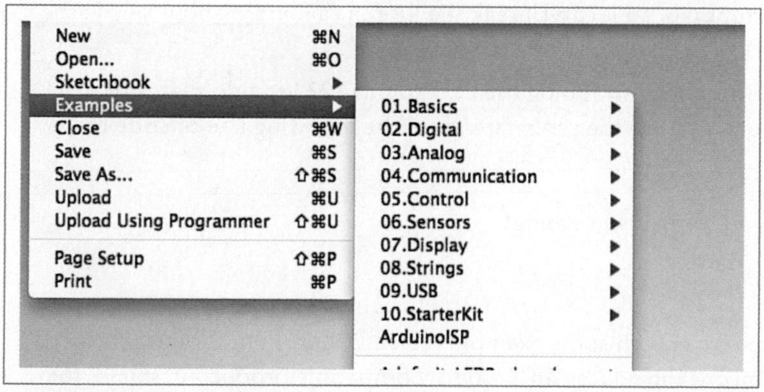

FIGURE 5.8 Need help learning how to program sketches? Look at Arduino's example code!

Okay, so you've opened up the Button sketch—now what? Let's play around with it.

Adapt the Code

The first step to learning how to code is to open a program that you *know* works, and then begin changing the elements one at a time to make the code conform to the needs of your project.

For example, let's change the pin numbers. Often, you can swap which jumper goes where. In the Button example, the button is on pin 2 and the LED on pin 13. What happens if you move the button to another pin? Let's change it to pin 10. On line 29 of the code, you see this:

```
const int buttonPin = 2;
```

Change the 2 to a 10 and upload the sketch. Voilà! You're finished, right? Give it a try.

> **NOTE**
>
> **Make Sure You're Wired**
>
> Of course, to properly test the sketch, you'll have to wire up the board as instructed in the sketch.

When you test it out, you'll discover that it doesn't work. What gives? It turns out that pin 2 was important. It's an interrupt pin, which has the ability to break into loop. As you might have noticed in the sketch, the loop is perpetually running, waiting to be told to turn on the LED. Pin 10 isn't an interrupt pin, so it can't break the loop. I talk more about this in the later section, "Interrupts."

Okay, that didn't work. How about swapping the LED to pin 12? Let's do that, but first change `buttonPin` back to 2. Change the pin reference in the following line of code to 12:

```
const int ledPin =  13;
```

Hooray! It worked, and you learned something!

Finding Example Code

Yes, the Arduino software comes with some example sketches, but more is out there to be found! Usually, when someone introduces an Arduino-compatible product—a shield, for instance—they courteously provide an actual sketch that they *know* works. This way, you can test out your board while learning about the syntax used to control it.

The most obvious way to find the example code for a specific product is to look up the product page on the seller's or manufacturer's website. Let's take the LoL (Lots of LEDs) Shield, which creates a 9×14 grid of LEDs that can be turned on and off individually, allowing for scrolling text and animations. Adafruit sells it (P/N 493) but doesn't manufacture it. Clicking on the Tutorials tab on its store website provides you with links to two pages on the creator's website. The first one shows you how to build the kit, and the other one provides example code to show you how to use it (see Figure 5.9).

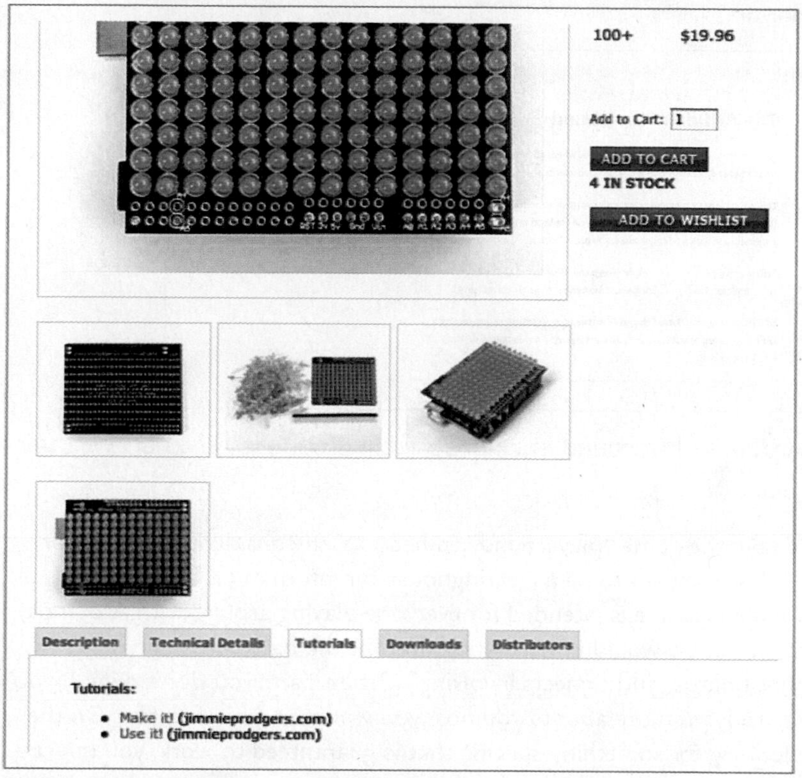

FIGURE 5.9 You can usually find example code on the website where you bought it.

However, what do you do if you are tackling your own project rather than buying someone else's kit? The following suggestions might help you find the code you need to get started.

Arduino Playground

If you can't find any example text, your next destination in hunting for code should be the Playground (http://playground.arduino.cc/), the Arduino platform's technical home (see Figure 5.10). It's packed with suggestions, tutorials, and code for nearly any situation you might encounter.

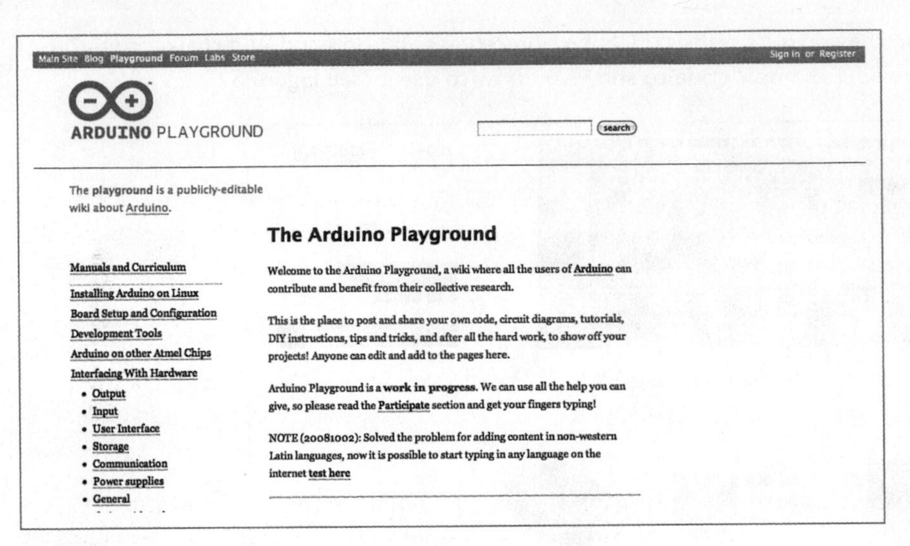

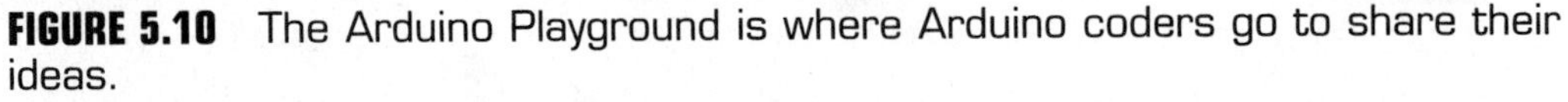

FIGURE 5.10 The Arduino Playground is where Arduino coders go to share their ideas.

That said, you should take the name "playground" to heart when considering looking for resources on the site. It's not meant to be a clearinghouse for information as much as, well, a playground. Because the resource is intended for everyone playing around with Arduinos, it has a lot of stuff you probably wouldn't need—bug fixes for the hardware, a guide to making your own circuit boards, and projects involving obscure parts you don't own. If you go there simply hoping to learn more about Arduinos, you won't be disappointed! On the other hand, if you're looking for something specific that is guaranteed to work, you might not find it.

If a beginner does want to get a sense of the materials found in the Playground, the best place to start might be the section marked "Manuals and Curriculum." It consists of numerous resources that can be downloaded to read offline, as well as links to third-party, beginner-level tutorials.

Libraries

Another source for example code are libraries, which are files of code used to do background work in sketches while keeping the actual sketch as neat and clean as possible. Libraries often include example sketches showing how to use the library. Want to learn more about how servo motors (for example) work? Download the Servo library (http://playground.arduino.cc/ComponentLib/servo) and install it to take advantage of the two sketches that come with it.

To learn more about libraries, including how to install them, see the section "All About Libraries," later in this chapter.

Sharing Example Code

The Arduino phenomenon is open source (see the nearby sidebar), which means that it subscribes to a philosophy where everything should be shared. For example, if you wanted to etch your own Arduino Uno circuit board and solder in all the components, you could totally do that—and some people do. The flip side of this sharing is that if you develop an open source project, you should tell people how you did it so they can learn from you.

It doesn't mean you're obligated to share, but it's the courteous thing to do if you used open source resources to create the project.

You have a couple of options for sharing code:

- **Arduino Playground**—Host your code on your own site, and then add a link to the Playground, which is a publically editable wiki, like Wikipedia. By hosting it on your own site, you can be sure that it won't simply vanish off the Playground. (It is publically editable, after all!) It also sends traffic to your site, enabling you to show off your cool projects!
- **Code.Google.com** or **GitHub.com**—These services offer repositories for storing and distributing code, so you don't need to host on your own site. Beyond that, the services offer some intriguing options, such as the ability for others to "fork" the project—that is, to spin off their own version of the project and take it to a different conclusion than you might have.

If you share code, feel free to include a comment like so:

```
// Code written by John Baichtal / nerdage.net
// If you reuse this code, please give attribution!
```

Furthermore, if you adapted someone else's code, you should include a reference in comments as well, giving credit where it's due!

WHAT IS OPEN SOURCE?

Open-source hardware (see Figure 5.11) and software are philosophies that espouse the total sharing of source code, schematics, designs, and so on. Let's pretend you're selling a robot kit. People can buy the kit from your web store, or they can simply download the files and build the project without sending you any money. That's right: You provide free downloads of everything a person needs to build the robot, including the Arduino code, the PCB schematics, and even the laser-cutter vectors for the robot's chassis.

FIGURE 5.11 Look for the open-source hardware logo. It looks kind of like a gear and has OSHW next to it.

So, one might ask, how is this a viable business practice? Why would someone buy your kit if they can download it for free? It turns out that it's considerably more expensive and labor-intensive to create one copy of a project than it is simply to buy a kit.

You would need to create the PCB, which involves either sending it to a service or etching it yourself. You would have to buy all the electronic components individually—kitmakers buy in bulk at a discount and pass their savings on to the customer. Finally, getting access to a laser to cut the chassis could be problematic or expensive—or both! Kitmakers really don't have anything to fear from individuals making their own. People who etch their own boards are unlikely to buy a kit anyway.

Open source hardware sellers do, however, have reason to fear certain shady electronics firms who take open source hardware (such as Arduinos) and have them manufactured at sweatshops and then sold at a discount as if they were the real thing. That said, you could certainly manufacture your own version of an open source project and then sell it—but you should make an effort to improve upon or customize it. No cloning!

Here are the generally agreed-upon rules of open source hardware:

- **Release the source**—This might seem obvious, but sometimes companies want the benefits of open source without actually giving it out. Do the right thing, and release the source even if it's not done.

- **Give credit where it's due**—If you used someone else's code to create your own sketch, mention it in comments. Really, try to mention anyone whose open source project contributed to yours.
- **If you're creating a project based on open source designs, make it open source**—Sharing in the generosity of the Arduino community without giving back is uncool.

If you still have your doubts about the commercial viability of open source hardware, consider two companies whose products I mention a lot in this book: Adafruit Industries and SparkFun Electronics. They're both million-dollar companies that deal exclusively in open source hardware.

To learn more about open source hardware, visit OSHWA, the Open Source Hardware Association at www.oshwa.org/.

More Functions and Syntax

Let's cover some of the functions and syntax you're likely to encounter while programming Arduino, besides those we've already covered. This section explores these only briefly, but I provide links to web pages where you can learn more.

Arithmetic

Arduinos are capable of handling math, and I'm not just talking about plus and minus. The math.h library, which manages the math functions built into the Arduino's microcontroller chip, is included with your Arduino environment and doesn't need to be downloaded. It can do all sorts of higher math, but here is some basic arithmetic:

```
z = x + y;
z = x - y;
z = x * y;
z = x / y;
```

If you need to do some higher math such as trigonometry, you can learn more about the math.h library on the Arduino website at www.arduino.cc/en/Math/H.

Arrays

Arrays are a way of managing a large amount of information. You can always tell an array function because it has brackets instead of parentheses, like this:

```
int myArray[10]={9,3,2,4,3,2,7,8,9,11};
```

The number in the brackets is the number of items in the array, and the items in the curly braces are the actual array. Each item is numbered, starting with 0—so the first item is myArray[0] and the eighth item is myArray[7].

To learn more about arrays, visit http://arduino.cc/en/Tutorial/Array.

Comparison Operators

Greater than, less than, equal to: These are comparison operators. You represent them in Arduino in a typical manner:

== (equal to)

!= (not equal to)

< (less than)

> (greater than)

<= (less than or equal to)

>= (greater than or equal to)

For

The For function repeats a step for a certain number of times, and then stops. For example, if you want to flash an LED 10 times, a For function is what you want, like so:

```
int led = 13;

void setup() {
 pinMode(led, OUTPUT);
 for (int item=0; item < 10; item++){
       digitalWrite(led, HIGH);
       delay(1000);
    Serial.print(item);

       digitalWrite(led, LOW);
       delay(1000);

    if (item==9){
       break;
}
  }
}

void loop() {
}
```

So, what's going on with that code? First of all, note that the For function is inside void setup, not void loop. This signifies that the For loop will run only once. If you put it in void loop, it will cycle from 0 through 9 infinitely.

Next, the integer item is examined by the loop. It starts at 0, and every time it loops, it increases by 1—that's the item++ reference. When the number reaches 9, the For loop breaks and the sketch moves on to void loop, which is currently empty.

To learn more about For loops, check out the Arduino tutorial at http://arduino.cc/en/Tutorial/ForLoop.

Include

The include reference indicates that another file, usually a library, will be included when the sketch uploads. For example, if you want to run a stepper motor using the stepper.h library, enter the following reference at the beginning of your sketch:

```
#include <Stepper.h>
```

See http://arduino.cc/en/Reference/Include for more information.

Increment/Decrement

An increment or decrement increases or decreases (respectively) the target number by one. You can see this in the For example earlier in this section. Do you remember where it said item++? The double plus increases the integer "item" by one. To decrease the integer by one, you would write item--. To see more about increments and decrements, see the following web page: http://arduino.cc/en/Reference/Increment.

Interrupts

What do you do if you want to break out or change the void loop part of the sketch? For example, look at the following modified Blink sketch. I changed it so you have to flick a switch for it to work.

```
int led = 13;

void setup() {
  pinMode(led, OUTPUT);
}

void loop() {
    int switch1 = digitalRead(2);
    if (switch1 == HIGH) {
  digitalWrite(led, HIGH);
  delay(1000);
  digitalWrite(led, LOW);
```

```
    delay(1000);
  }
}
```

The key ingredient here is that switch1 is connected to pin 2 on the Arduino. Two interrupt pins are on the Arduino Uno, digital pins 2 and 3. If you want to modify or stop a loop, you must use those pins.

If/Else

This function enables you to set up conditions that, if met, will trigger an action and, if not met, will trigger a different action. You can see the `if` part in the preceding sketch, where pressing a button enables an LED to blink on and off. But where does the `else` fit in? The `else` event is triggered if the `if` statement is false. For example, if you had two LEDs in the preceding sketch, you could add an else to turn on a second LED if the switch is not thrown, like so:

```
switch1 == HIGH) {
  digitalWrite(led1, HIGH);
}
else
{
  digitalWrite(led2, HIGH);
}
```

To learn more about if/else functions, check out http://arduino.cc/en/Reference/else.

Mapping

The mapping function remaps a number from one range to another. For example, if you get a reading of 0 through 1023 on an analog sensor, you could remap it so it instead returns a reading of 5 through 23. Here is the syntax:

```
mySensor = map(mySensor, 0, 1023, 5, 23);
```

You get it! See http://arduino.cc/en/Reference/map for more information.

Random

If you want to generate a pseudo-random number, use this simple function:

```
randomNumber = random(min, max);
```

The `min` and `max` refer to the minimum and maximum values possible. So, `random(5,10);` would return a pseudo-random number between 5 and 10. But wait, what is a pseudo-random number? It turns out that computers, including Arduinos, are so logical and orderly that they simply cannot create a random number. Instead, they employ such tricks as counting the number of milliseconds since the device was turned on. Although this is not *truly* a random number, it's close enough for most of us.

Switch/Case

Another use of the `if` statement is to switch between a number of options depending on the valuable of the variable. Suppose `myVariable` can be either A, B, or C. You could set it to trigger an action depending on the result. In this example, the Arduino prints the name of a fruit:

```
switch (myVariable) {
    case A:
      Serial.print("Apple");
break;
    case B:
      Serial.print("Banana");
      break;
    case C:
      Serial.print("Cherry");
      break;
}
```

To learn more, see the following web page: http://arduino.cc/en/Reference/SwitchCase.

While

The `while` function creates a loop that runs indefinitely until the conditions are met. In the following code snippet, the `while` loop runs until a button is pressed 100 times:

```
buttonPress = 0;
while(buttonPress < 100){
  buttonpress++;
}
```

There's a great `While` tutorial on the Arduino site: http://arduino.cc/en/Tutorial/WhileLoop.

Debugging Using the Serial Monitor

The easiest way to debug a sketch that successfully uploads to the Arduino but nevertheless doesn't work correctly is to observe it in the serial monitor. The serial monitor is a window that displays the serial data traffic going to and from the Arduino. The way it works is that while connected to the computer via the USB cable, the Arduino environment receives info from the Arduino as directed by the sketch, and you can send data back to the Arduino the same way.

Let's go over the features of the serial monitor, shown in Figure 5.12.

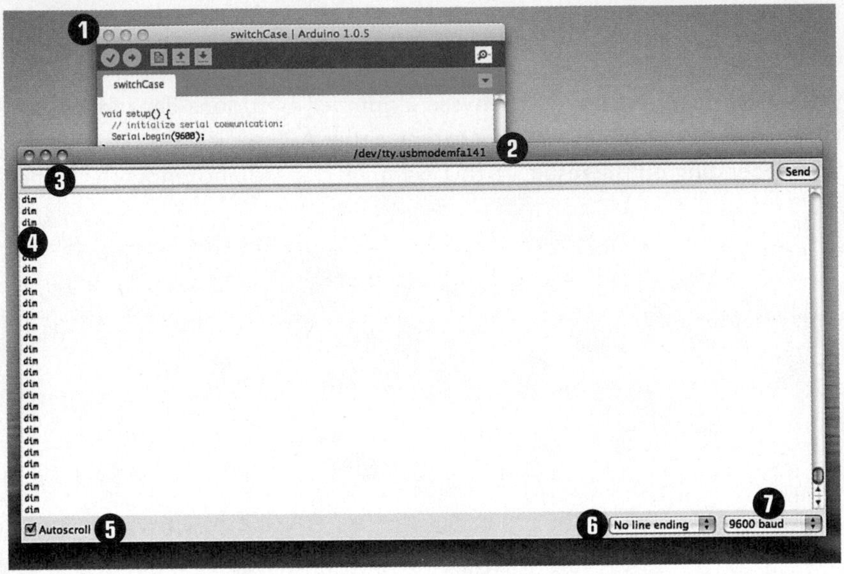

FIGURE 5.12 The serial monitor displays data from an Arduino running the switchCase example sketch.

1. **The sketch window**—The serial monitor needs a sketch window open to launch, but it doesn't have to be the actual sketch you're interacting with.
2. **Port number**—This is not something you're likely to worry about or have to tinker with.
3. **Text entry field**—Do you want to send text to your Arduino via the serial monitor? Just type in your text and click Send. Note that the sketch must have the correct functions in place to do anything with this text!
4.  **Display area**—The text from the serial monitor displays here.
5. **Autoscroll**—This defaults to "checked" and automatically scrolls the page to display the latest lines of information from the Arduino. Unchecking this means that you'll have to use the window's scrollbars to navigate down to the end of the page to see the new lines.
6.  **Line ending**—This pulldown menu gives you options for terminating lines of text. It defaults to "no line ending."
7. **Baud rate**—This is the speed of communication between the computer and the Arduino. You set the speed in the serial monitor using this drop-down menu. You set the speed in the Arduino using code, as described next.

More intriguingly, you can use these notes to tell where a bug is in your sketch. One way to do this is to sprinkle your code with functions that send text via serial. Consider the AnalogReadSerial sketch, one of the examples included with the Arduino software. First, look at the function within `void setup()`:

```
void setup() {
  // initialize serial communication at 9600 bits per second:
  Serial.begin(9600);
}
```

`Serial.begin()` turns on serial communication between the computer and Arduino, and the number refers to the number of bits per second. The number 9600 is a pretty common standard but you don't have to pick a particular speed—just make sure to match the speed in your serial monitor. Next, look at the loop:

```
void loop() {
  // read the input on analog pin 0:
  int sensorValue = analogRead(A0);
  // print out the value you read:
  Serial.println(sensorValue);
  delay(1);        // delay in between reads for stability
}
```

Look at the `Serial.println()` function. It sends information—in this case, the reading from analog pin 0—to the serial monitor. What is cool about it is this: If you don't see the reading, then you'll know that something went wrong in the sketch and approximately where it went wrong. If you had a hundred-line sketch with those `Serial.println()` functions scattered all over, you could literally follow along with the sketch in the monitor and could pinpoint exactly where the problem might be. Try it!

All About Libraries

Another resource the Playground offers is a selection of libraries. These consist of functions collected in a separate document, so you can reference them from your main script while keeping the code clean.

Pretty much any time you have a sensor or other electronic device connected to the Arduino, you'll need code to control it, and a library is often the best way to manage that code. If you buy a new component, take the time to do a search for a library—it might save you a bunch of time!

Okay, go ahead and open a library. Let's use the library Servo.h, which you can find at http://www.arduino.cc/en/Reference/Servo. It helps sketches control servo motors. The following is an example of how it works. There is a basic sketch, Sweep, which can be used to control servos. Right off the bat you can see that it needs a library to work:

```
#include <Servo.h>
```

This reference serves as a notice that the library Servo.h is needed to run the sketch, and if Servo.h is located along with all of your libraries, the relevant data loads automatically so that the sketch operates as expected. If the library is missing, the sketch sends an error message when you try to upload to the Arduino.

Next, look at a single instance of the library in use. In the Sweep sketch, a new servo is declared:

```
Servo myservo;  // create servo object to control a servo
```

The servo object (called `myservo`) is created as described by the library:

```
class Servo
{
  private:
    uint8_t _index;
    uint8_t _pin;
    uint16_t _duty;
    static uint8_t _count;
    static Servo* _servos[];
    static int8_t _current;
    static uint16_t _positionTicks;
    static void start();
    static void end();
    static void service();
  public:
    Servo();
    uint8_t attach(int);
    void detach();
    void write(int);
    uint8_t read();
    uint8_t attached();
};
```

Private refers to functions used by the library itself, often to do background work such as assigning bytes. The *public* functions are the ones you reference in the actual sketch.

As your code gets more and more complex, you might get the idea that you could benefit from building your own library. Although discussing this task falls outside the scope of this book, Arduino has a fine tutorial on its site at http://arduino.cc/en/Hacking/LibraryTutorial.

Resources for Learning Programming

I could easily write 10 pages of Arduino resources, but the following sections introduce just a few of the coolest.

Books

- *Arduino Adventures: Escape from Gemini Station* (Apress 2013, ISBN 978-1430246053) is a kids' book by James Floyd Kelly and Harold Timmis that combines cool Arduino projects with a fun Young Adult fiction storyline.
- *Arduino Cookbook* (O'Reilly 2011, ISBN 978-1449313876) by Michael Margolis is considered by many to be the definitive Arduino reference. Do you need to know how to run a seven-segment display using a breadboarded LED driver? Look it up in the Cookbook.
- *Getting Started with Arduino* (O'Reilly 2011, ISBN 978-1449309879) is a pocket-sized beginner's guide. It is written by Massimo Banzi, one of the founders of Arduino, so you know you're getting authoritative information!
- *Make: Lego & Arduino Projects* (Make 2012, ISBN 978-1449321062) by John Baichtal, Adam Wolf, and Matthew Beckler. The book focuses on using Arduinos to control Lego Mindstorms robots.
- *Making Things Move* (TAB Electronics 2011, ISBN 978-0071741675) by Dustyn Roberts. This is not an Arduino book exactly, but it packs tons of information about electronics and robotics.

Websites

- Adafruit Industries (adafruit.com) is one of the main sites for DIY electronics. It features tutorials and code, as well as an excellent store packed with the stuff you need to do Arduino projects.
- Arduino Playground (playground.arduino.cc) was mentioned earlier in the chapter. It's the motherlode of Arduino code, both finished as well as experimental.
- Instructables (instructables.com) is chock-full of DIY tutorials, and many of them cover electronics and Arduino topics.
- Make (makezine.com) is the granddaddy of the modern DIY scene. Make puts on Maker Faires (makerfaire.com) and publishes a paper magazine.
- SparkFun (sparkfun.com) is like Adafruit: a cool electronics store with associated tutorials, videos, and other resources.

The Next Chapter

In Chapter 6, "Sensing the World," we continue our exploration of sensors, which take readings from the surrounding environment and transmit the info to the Arduino. In it, you'll learn how to build a mood lamp that takes those readings and changes the color of an LED to reflect the environment around it!

Sensing the World

One of the most powerful tools you can plug into your Arduino is a *sensor*, a small electronic device that enables the microcontroller to take readings from its surroundings, reacting in accordance to its program. For example, you could program the Arduino to turn on a fan when a temperature reaches a certain level or to turn off a lamp when the sun comes up.

In Chapter 1, "Arduino Cram Session," you learned a little about some sensors —ultrasonic, temperature, flex, and light. However, many more types are available! In this chapter, you learn more about additional sensors and how to control them, and then you use that knowledge to build a sensor-controlled project—a mood light that changes its color depending on environmental conditions (see Figure 6.1).

FIGURE 6.1 Learn how to make a cool mood light in this chapter.

Lesson: Sensors

What exactly does a sensor do? We know about temperature and light sensors, but no sensor exists that is as sophisticated and intuitive as human sight, for example, and there's no such thing as a "scent sensor" that triggers an action when someone cooks up a hot dog nearby. But what can sensors do?

Sensors are fairly simplistic—they usually focus on measuring one phenomenon. For instance, a passive infrared sensor (described in Chapter 2, "Breadboarding") detects abrupt changes in temperature in its field of view, and translates that as movement. A barometric sensor measures air pressure much the same way a barometer does, and sends the reading to an Arduino. A photo resistor limits the flow of current with a value depending on how much ambient light the resistor detects. Knowing how a sensor works helps you use it effectively in your project.

Figure 6.2 shows an accelerometer, which is a sensor that determines what direction an object is traveling and its speed.

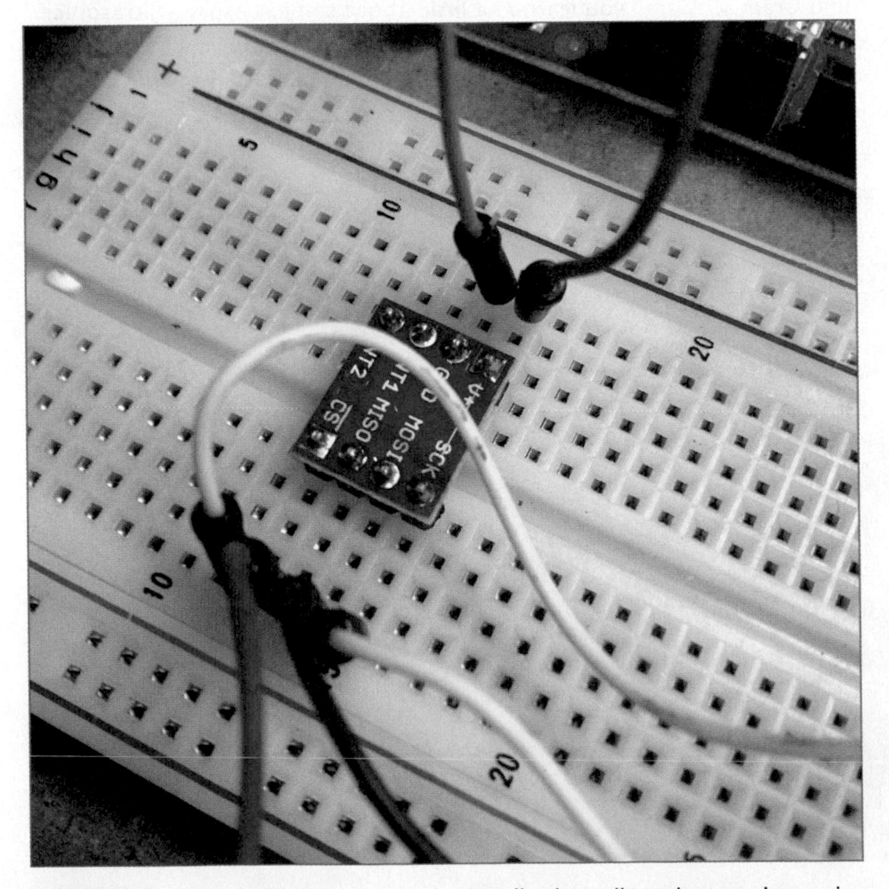

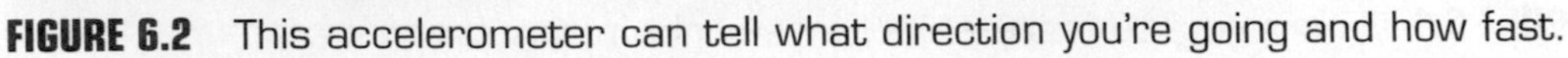

FIGURE 6.2 This accelerometer can tell what direction you're going and how fast.

Speaking broadly, the two categories of sensor are digital and analog. Let's explore the differences.

Digital Versus Analog

Digital and *analog* are two methods of transmitting information. The Arduino world uses both methods a lot, so you'll need to know both.

Digital

Digital data consists exclusively of 0s and 1s (see Figure 6.3). An example of a digital sensor is an accelerometer, which sends a series of data points (speed, direction, and so on) to the Arduino. Usually digital sensors need a chip in the sensor to interpret the physical data.

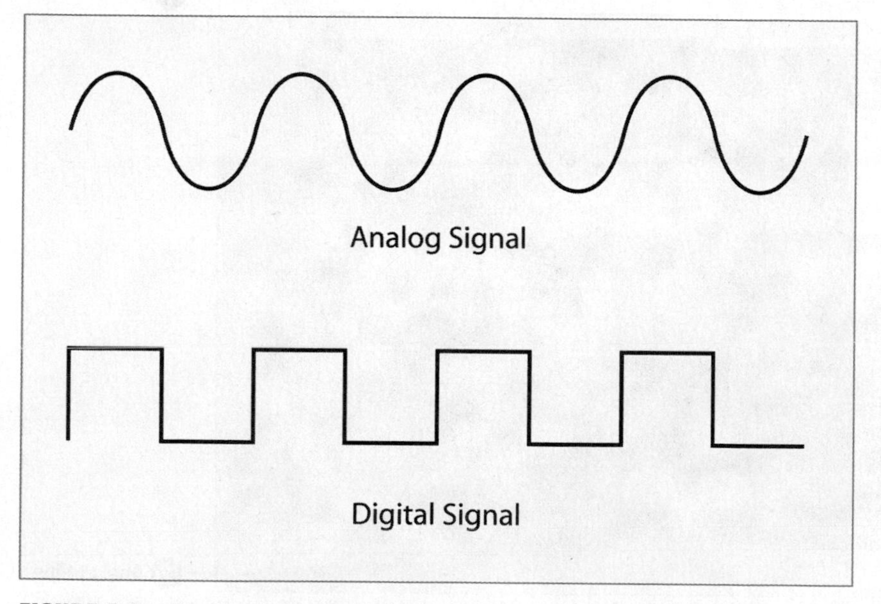

FIGURE 6.3 Digital information is transmitted with a series of 1s and 0s; analog data consists of a modulated signal.

Because digital sensors send only data, they need a microcontroller to interpret the readings. A digital light sensor, therefore, could not be used as a photo resistor, although you could rig a microcontroller-controlled circuit to do the same thing.

Analog

Analog data is transmitted as a continuous signal, almost like a wave. In other words, an analog sensor doesn't send a burst of 1s and 0s like digital sensors do; instead, the sensor modulates a continuous signal to transmit data.

You can use analog sensors in circuits without needing a microcontroller to interpret the signal. In the case of the photo resistor, you could actually use an analog light sensor as a photo resistor.

Connecting Digital and Analog Sensors

Not surprisingly, the Arduino reserves some pins for digital input and output, and others for analog. A servo's data wire plugs into a digital pin, whereas an analog light sensor sends its data reading to an analog pin.

Which pins are which? You can easily tell just by looking at the Arduino. The digital pins are all along one side (on the top of Figure 6.4), and the analog pins are on the opposite side—the bottom right of the figure. If you forget, just look at the printing on the board itself.

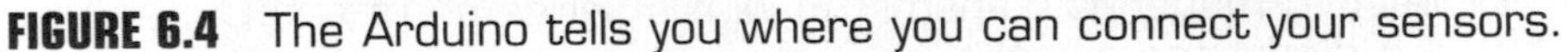
FIGURE 6.4 The Arduino tells you where you can connect your sensors.

Additionally, when programming the Arduino sketch—which is what the Arduino world calls programs—you'll need to declare your pins. I cover how to do that in Chapter 5, "Programming Arduino."

Know Your Sensors

What exactly is a sensor? On a certain level, it's a device that sends information to the Arduino based on some external factor the device measures. For instance, the barometric sensor (described shortly) measures air pressure and returns a reading based on what the sensor detects.

The following sections describe some of the sensors you can use with your Arduino projects.

Accelerometer

Figure 6.5 shows the ADXL362 accelerometer as part of a small circuit board sold by SparkFun (P/N SEN-11446). It can tell in what direction it's going and how fast; it sends the data digitally. You can make fun projects with the accelerometer, such as a self-balancing robot, which uses an accelerometer to tell when it's tipping over, and then moves to balance itself.

FIGURE 6.5 SparkFun's ADXL362 lets you plug in an accelerometer to a breadboard.

Barometric

The staple of science-fair weather stations, the barometric sensor is basically a digital barometer hooked up to an Arduino. The sensor shown in Figure 6.6 (Adafruit P/N 391) monitors air pressure and sends readings to the Arduino using I²C, a method of transmitting data along a single wire.

FIGURE 6.6 The BMP085 barometric sensor monitors air pressure. Credit: Adafruit Industries

Encoder

The servo motor shown in Figure 6.7 is equipped with a rotation sensor called an encoder. When the motor's hub turns, the encoder sends data back to the Arduino describing the precise angle of rotation. One possible use for this would be to create a knob that controls a servo—the Arduino reads in the position of the knob's encoder and instructs the servo to respond.

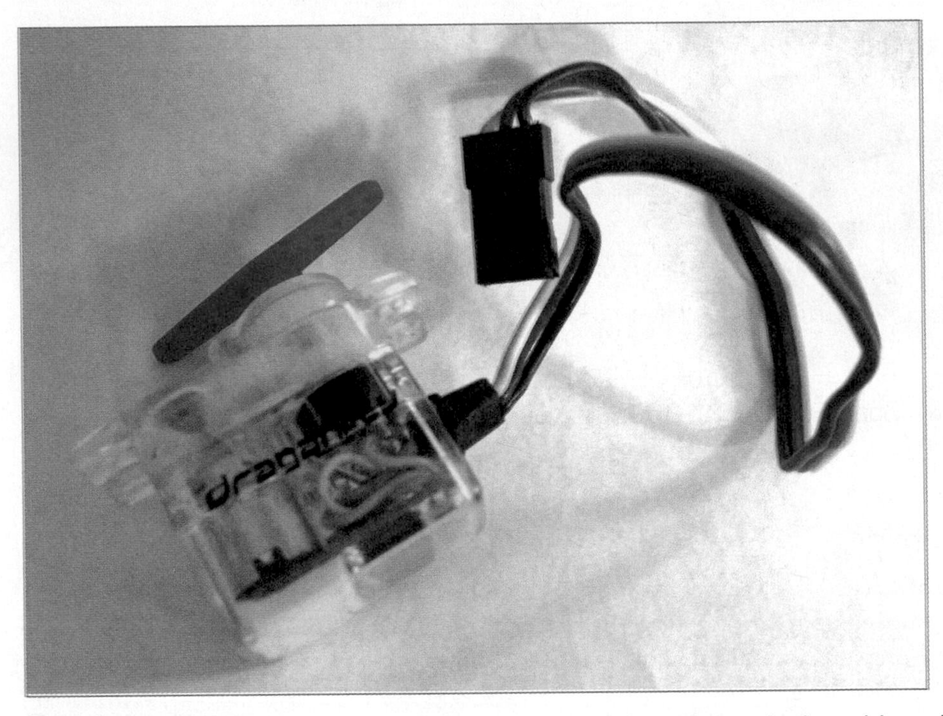

FIGURE 6.7 A hobbyist servo motor transmits data through the white wire.

Gas

The MQ-4 methane gas sensor shown in Figure 6.8 (Sparkfun P/N SEN-09404) is often used to make gas-leak alarms, as well as (ahem) fart detectors.

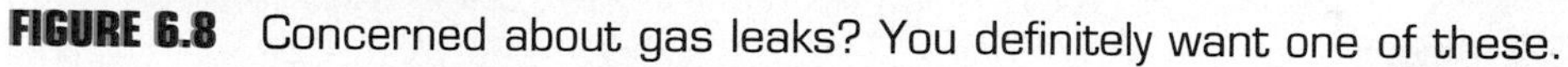

FIGURE 6.8 Concerned about gas leaks? You definitely want one of these.

Hall Effect

The Hall Effect sensor shown in Figure 6.9 detects the presence of magnets nearby. This is great for activating another circuit without needing to be in physical contact—for example, if the sensor is separate from a circuit by a sheet of glass. It's also great for checking the proximity of another object that has a magnet embedded in it.

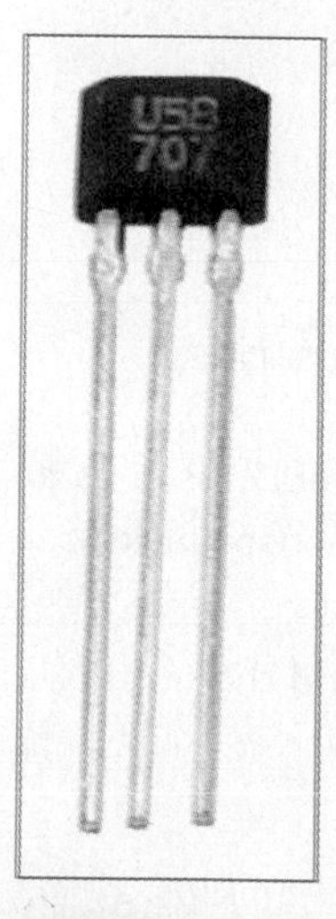

FIGURE 6.9 The Hall Effect sensor produces voltage when magnetic fields are nearby.

Infrared

Most television remote controls use coded pulses of infrared (IR) light to tell the appliance what you want it to do. For instance, "turn off" might be a certain code, whereas "lower volume" might be another. The television has one of these sensors to receive the IR.

IR sensors (see Figure 6.10, Adafruit P/N 157) can be used for the same reasons one might use a photo resistor or light sensor, except that because it uses IR, you won't see any annoying flashes of light. This characteristic makes IR sensors useful for projects where you don't want bright LEDs shining everywhere, or if you're concerned with "garbage light" incorrectly triggering the sensor.

FIGURE 6.10 This sensor detects infrared light. Credit: Adafruit Industries.

For example, you could position an IR sensor on a robot, right next to an IR LED, used as a proximity sensor. When the light of the LED reflects off a nearby object, the sensor picks it up and sends a signal to the Arduino.

Most IR sensors look like small black boxes with a bulb-shaped protrusion and three leads.

Piezo Buzzer (Knock Sensor)

A funny thing about some electronic components is that they work in reverse! Shining a light on an LED generates a tiny trickle of voltage. A piezo buzzer—often called a knock sensor—works the same way (see Figure 6.11). If you send voltage through a piezo, it vibrates and makes a buzzing noise. Similarly, if you vibrate the piezo manually (for example, by tapping on it) a small amount of voltage is generated. This means you can also use it as a vibration sensor!

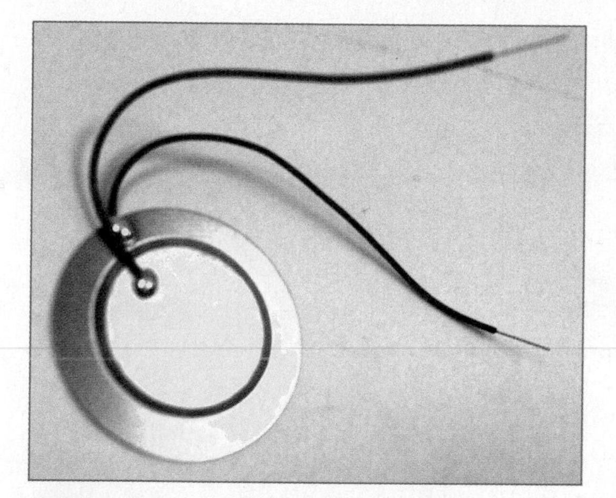

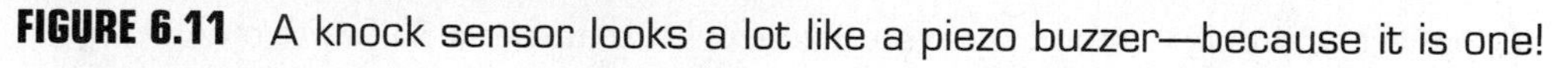

FIGURE 6.11 A knock sensor looks a lot like a piezo buzzer—because it is one!

Sound Sensors

A sound sensor (see Figure 6.12) is essentially a microphone that picks up nearby sound vibrations. Those vibrations generate tiny voltages, which are picked up by the sensors. You often use an amplifier to magnify those signals to play on a speaker. However, you can also use the sensor to trigger actions in an Arduino program without amplification.

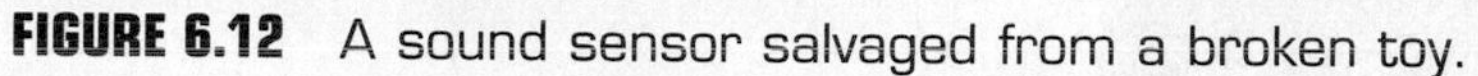

FIGURE 6.12 A sound sensor salvaged from a broken toy.

Tilt Sensors

A tilt sensor, like the one shown in Figure 6.13 (Adafruit P/N 173), is a tube with a ball rolling around in it. When the sensor is in its normal vertical position, the ball connects two wires, but if the sensor tips and the ball rolls away from the contacts, the connection is lost. For this reason, the sensor is sometimes called the "poor man's accelerometer."

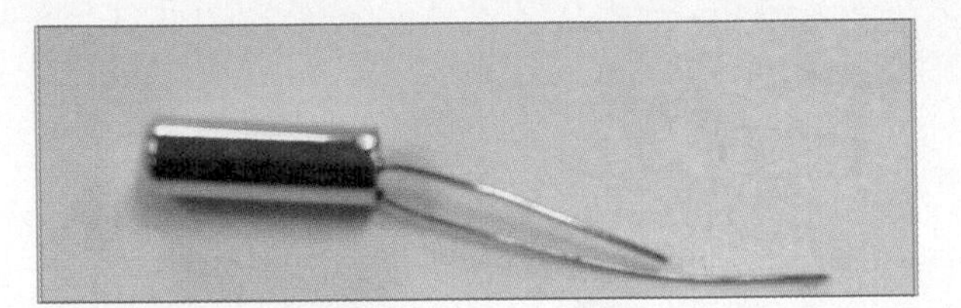

FIGURE 6.13 A tilt sensor can tell when it's being inverted.

Project: Mood Light

In this project, you build a mood light (see Figure 6.14) that uses a ShiftBrite module, a very bright RGB LED mounted on a circuit board, useful for just this sort of application. Controlling the LED is an Arduino Uno (of course!) connected to a small solar panel used as a light sensor, a temperature and humidity sensor, and a small microphone serving as a sound sensor. For an enclosure, you explore the world of kerf-bending, a technique where you cut slits into a panel of wood, which allows it to curve and flex without breaking.

FIGURE 6.14 The temperature and humidity sensor (the white plastic module) and the sound sensor (the little button in the middle of the picture) both help control the color of the lamp.

PARTS LIST

You'll need the following parts to build your mood light:

- Arduino Uno
- ShiftBrite module: See the nearby note for more information (SparkFun P/N 10075).
- Light sensor: I used a mini solar panel similar to Jameco P/N 2136913, but you could use any light sensor or solar panel.
- DHT22 temperature and humidity sensor (Adafruit P/N 393)
- Electret microphone (Adafruit P/N 1064) used as a sound sensor
- Some sort of lampshade for the ShiftBrite; I used the glass globe from a lawn light (Hampton Bay SKU #708 407).
- 6 #4-40 x 1" bolts with nuts
- 4 3/8" plastic standoffs (SparkFun P/N 10461)
- Mini breadboard
- Assorted jumpers
- A sheet of MDF or fiberboard—an 18-inch × 24-inch sheet should be plenty.

NOTE

The ShiftBrite Module

The ShiftBrite is an LED module created by Garrett Mace that enables Arduino fans to control a high-brightness RGB LED very precisely (see Figure 6.15). RGB represents the colors Red, Blue, and Green, which combined can display any color of light. The ShiftBrite, therefore, has three elements—one for each color—and the Arduino controls the values of all three.

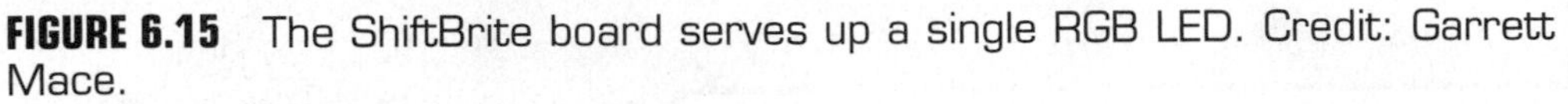

FIGURE 6.15 The ShiftBrite board serves up a single RGB LED. Credit: Garrett Mace.

It's kind of cool how the Arduino controls the ShiftBrite's color and brightness. It does so with serial communication, which you used in Chapter 4, "Setting Up Wireless Connections." Ordinarily, an Arduino controls an LED by changing the voltage the LED receives. In a practical sense, this makes controlling large numbers of LEDs difficult because the Arduino has only so many pins.

The solution is to use an LED driver, a microchip that controls multiple LEDs and takes its orders from an Arduino via a serial connection, where digital components like ShiftBrites can be controlled with just a few wires. Guess what? The A6281, a tiny LED driver, controls the ShiftBrite's LED. No only does this allow you to use data to change the LED's color and brightness, but it enables you to connect multiple ShiftBrites together in a string, all controlled by the same number of data pins (four) as it would take to control a single module.

You can learn more about the ShiftBrite on the module's product page: http://macetech.com/blog/node/54.

Instructions

Let's build it! The mood lamp is a relatively simple build, but there are a couple of complicated steps:

1. **Cut and assemble the enclosure**—I used 5mm fiberboard to cut out shapes on the laser cutter, using a technique called *kerf bending* (see "Alt.Project: Kerf-bending" at the end of this chapter). You can download the file I used from https://github.com/n1/Arduino-

For-Beginners. Alternatively, you can design your own enclosure! Figure 6.16 shows the enclosure walls that have been cut out on a laser cutter using kerf bending. I cut a big hole for the power supply, and the notches along the upper edges are for attaching the top.

FIGURE 6.16 The project enclosure uses a kind of cutting called kerf bending, allowing the wood to curve around to form the walls.

2. **Build the LED platform**—Next, add the LED assembly. I figured out how to attach the globe to the laser-cut box—see step 5. Basically, this involves a small panel that fits into the globe and is tensioned with a pair of screws (see Figure 6.17).

FIGURE 6.17 The ShiftBrite sits on a platform in the globe, using a mini breadboard to manage the wires.

3. **Attach the ShiftBrite**—To attach the ShiftBrite, stick a mini breadboard onto the panel using its adhesive backing or hot glue. Plug in the ShiftBrite, and then connect the wires to the correct holes on the breadboard, as shown in Step 4.
4. **Make the ShiftBrite wire connections**—Wiring up the ShiftBrite looks intimidating but don't be fazed! It's actually quite easy. It needs six wires:
   - V+ on the ShiftBrite plugs into the port marked 5V on the Arduino.
   - DI on the ShiftBrite plugs into port 10.
   - LI plugs into port 11.
   - EI plugs into port 12.
   - CI plugs into port 13.
   - GND plugs into GND.
5. **Attach the globe**—I used one from a yard light I bought at Home Depot (see Figure 6.18) and I've seen them at a lot of stores. I disassembled the light and set aside everything but the globe. To attach the globe, I drilled two screw holes in the top panel of the box, plus a larger hole for the wires. I used a small piece of wood that fits into the globe but also attaches to the top panel of the box, as shown previously in Figure 6.17. Tightening the bolts secures the globe to the top.

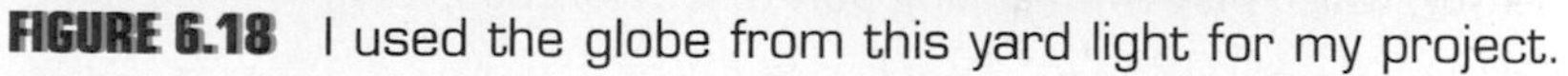

FIGURE 6.18 I used the globe from this yard light for my project.

6. **Connect the Arduino to the box**—After you attach the globe and wire up the ShiftBrite, you attach and wire up the Arduino. I connected the Arduino to the box using #4 bolts threaded through holes I drilled in the bottom panel, with 3/8" plastic standoffs. See Figure 6.19.

FIGURE 6.19 The Arduino, all wired up!

7. **Wire up the sensors**—See Figure 6.20. The light sensor plugs into analog port 1 and GND. The temperature sensor plugs into 3V3 and analog port 3, and shares a ground with the sound sensor, which plugs into analog port 2. It's not too tricky!

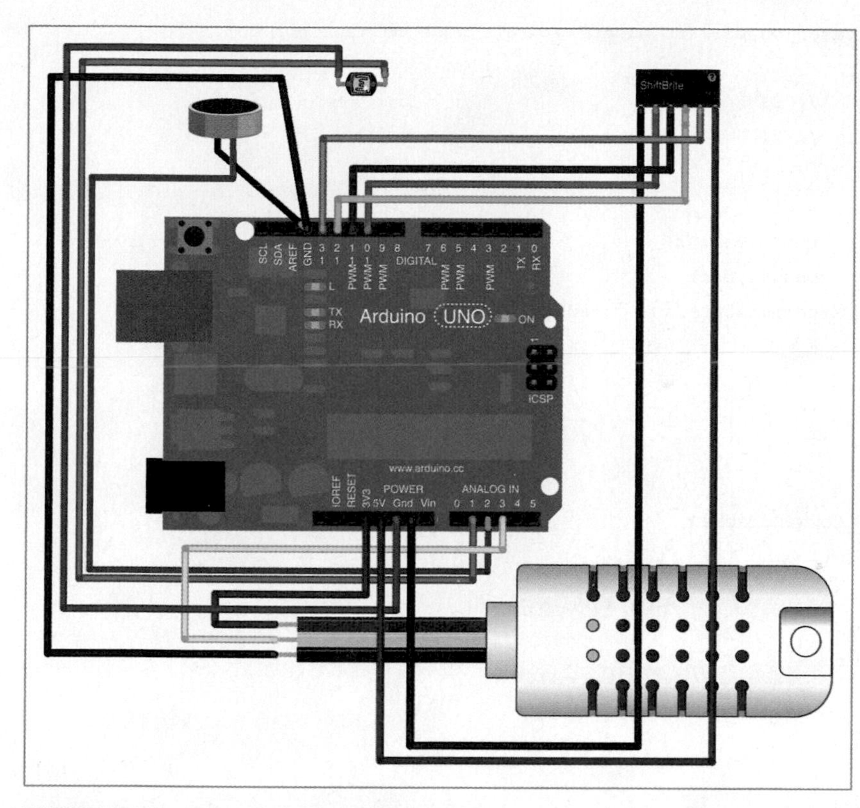

FIGURE 6.20 That's a lot of wires!

Mood Lamp Code

Here is the Arduino code for the mood lamp. Naturally, you'll want to tweak the code depending on your environmental conditions. For instance, a bright room might require a higher threshold for display; I show you how to in the sketch—look where it says, "adjust tolerances here." All you have to do is change the number by that notation to change how the Mood Lamp reacts to that environmental condition.

Finally, if you don't remember how to upload a sketch to your Arduino, refer to Chapter 5.

NOTE

Code Available for Download

You don't have to enter all of this code by hand. Simply go to https://github.com/n1/Arduino-For-Beginners to download the free code.

```
// This code is based on Garrett Mace's example code on macetech.com.

int datapin  = 10; // DI
int latchpin = 11; // LI
int enablepin = 12; // EI
int clockpin = 13; // CI
int pspin = A1; // photo sensor
int sspin = A2; //sound sensor
int thspin = A3; //temperature and humidity sensor

int light = 0;
int sound = 0;
int temp = 0;

unsigned long SB_CommandPacket;
int SB_CommandMode;
int SB_BlueCommand;
int SB_RedCommand;
int SB_GreenCommand;

void setup() {
   pinMode(datapin, OUTPUT);
   pinMode(latchpin, OUTPUT);
   pinMode(enablepin, OUTPUT);
   pinMode(clockpin, OUTPUT);
   pinMode(pspin, INPUT);
   pinMode(sspin, INPUT);

   digitalWrite(latchpin, LOW);
   digitalWrite(enablepin, LOW);

     Serial.begin(115200);            //  setup serial


}

void SB_SendPacket() {
   SB_CommandPacket = SB_CommandMode & B11;
   SB_CommandPacket = (SB_CommandPacket << 10)  | (SB_BlueCommand & 1023);
   SB_CommandPacket = (SB_CommandPacket << 10)  | (SB_RedCommand & 1023);
   SB_CommandPacket = (SB_CommandPacket << 10)  | (SB_GreenCommand & 1023);
```

```
    shiftOut(datapin, clockpin, MSBFIRST, SB_CommandPacket >> 24);
    shiftOut(datapin, clockpin, MSBFIRST, SB_CommandPacket >> 16);
    shiftOut(datapin, clockpin, MSBFIRST, SB_CommandPacket >> 8);
    shiftOut(datapin, clockpin, MSBFIRST, SB_CommandPacket);

    delay(1);
    digitalWrite(latchpin,HIGH);
    delay(1);
    digitalWrite(latchpin,LOW);
}

void loop() {

  light = analogRead(pspin) + 50; // adjust tolerences here
  sound = analogRead(sspin) + 50; // adjust tolerences here
  temp = analogRead(thspin) / 50; // adjust tolerences here

    SB_CommandMode = B01; // Write to current control registers
    SB_RedCommand = 127; // Full current
    SB_GreenCommand = 127; // Full current
    SB_BlueCommand = 127; // Full current
    SB_SendPacket();

    delay(2500);

    SB_CommandMode = B00; // Write to PWM control registers
    SB_RedCommand = sound;
    SB_GreenCommand = light;
    SB_BlueCommand = temp;
    SB_SendPacket();

        Serial.println("Light: ");
         Serial.println(light);
        Serial.println("Sound: ");
         Serial.println(sound);
        Serial.println("Temp: ");
         Serial.println(temp);
}
```

Alt.Project: Kerf Bending

Kerf bending is a clever way of bending wood. It's best done with the help of a laser cutter, although using saws to get the same effect is possible. The technique involves making a series of cuts in the material close together, in effect making the wood thin and flexible in spots. One obvious way to use this technique is to make a box by using kerf bends for the corners (see Figure 6.21).

FIGURE 6.21 Laser-cut slits (circled here) in this fiberboard allow it to flex.

If you want to create a laser-cut box design, you have a couple of options:

- You can download a box design from Thingiverse.com. This is a resource for people using 3D printers and laser cutters, and all the designs can be adapted. For instance, the box design I use in this chapter is a derivation of SNIJLAB's Folding Wood Booklet design—Thing #12707.
- Another option is the service at MakerCase.com. This site lets you type in the dimensions you want as well as select the particulars of your material such as thickness and connector gauge, and then the service generates the laser files for you. It's slick!

Both of these options are free, other than the actual cost to mill the designs.

The Next Chapter

In Chapter 7, "Controlling Liquid," you'll learn how to—wait for it!—use an Arduino to start and stop the flow of liquid. You'll then build a cool LEGO plant-watering robot with the skills you learn.

7

Controlling Liquid

Water and electronics don't mix. Well, mostly. But in this chapter, you explore three different ways of controlling liquids using electricity:

- **Pressurized reservoir**—This is a container of liquid that is pressurized with an air pump, forcing liquid out of an exit tube (see Figure 7.1).
- **Peristaltic pump**—A peristaltic pump massages a tube, forcing the liquid through while never actually touching it.
- **Solenoid valve**—A solenoid valve is an electrically controlled valve, opening up when it receives the proper voltage.

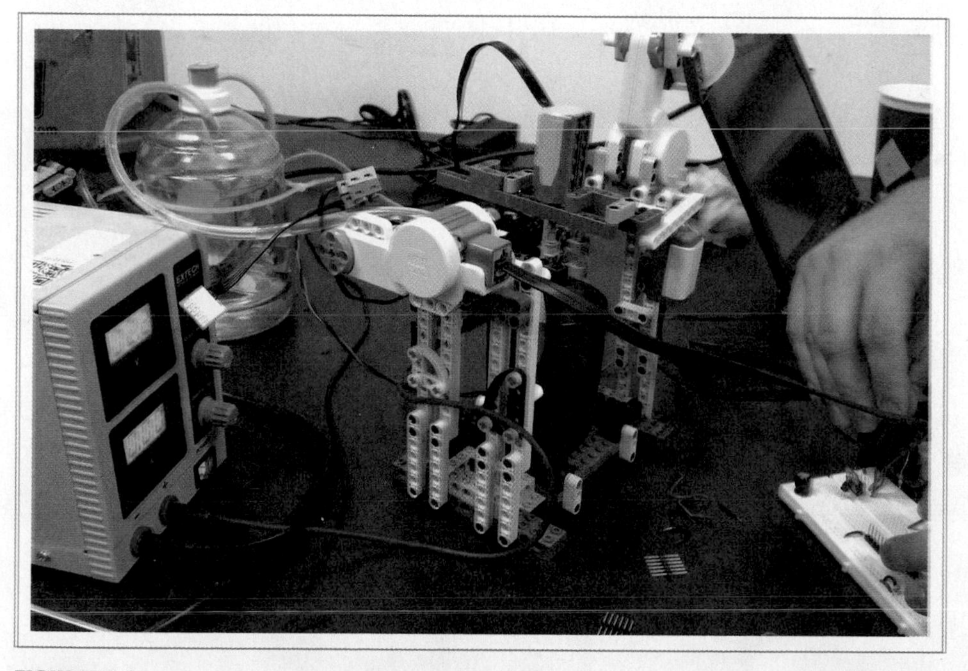

FIGURE 7.1 This LEGO chocolate milk–making robot uses a pressurized reservoir (the green jug) to pump milk.

After exploring these techniques—all of which can be controlled with an Arduino—you apply what you've learned and move on to the project: a robot that is programmed to water your plants on a schedule.

Lesson: Controlling the Flow of Liquid

Let's go over three ways to start and stop the flow of liquid.

Solenoid Valve

Let's begin with this solenoid valve from Adafruit (P/N 997) that can be triggered with a signal from your Arduino (see Figure 7.2). The valve is ordinarily closed, but when you apply 12 volts, the solenoid opens the valve and water goes through. When the voltage stops, the valve closes and the water stops. Yay, technology! If you remember from Chapter 1, "Arduino Cram Session," a solenoid is similar to a motor except instead of rotating the shaft, it pushes and pulls it. In the case of this valve, the solenoid opens and closes the valve.

FIGURE 7.2 A solenoid valve opens when it receives the correct voltage.

The valve connects to the water supply with half-inch plastic piping, automatically giving you an ecosystem of plumber's tubing and connectors you can buy in most hardware stores. I ended up using this valve for a plant-watering robot, precisely because of this flexibility. This convenience disguises one cool benefit—you can connect your robot directly to the main water supply rather than relying on a container of liquid.

If you end up using one of these valves, you should remember that the valve requires around 3 psi of water pressure to work properly, and only works in one direction.

Pressurized Reservoir

One disadvantage of the solenoid valve is that it's not food safe (see the later caution, "Food Safety"). This means you shouldn't use it for handling food and drink for human consumption. A pressurized reservoir avoids this problem by having air pressure move the liquid. One common use for a pressurized reservoir is a squirt gun (see Figure 7.3).

FIGURE 7.3 Does a pressurized reservoir sound familiar? It should! It's the method many squirt guns employ to shoot water.

Here's how a pressurized reservoir works:

1. A closed container of liquid is equipped with two tubes—one positioned just above the liquid level, and the other one all the way down to the bottom of the container.
2. When air comes in the top tube, it pressurizes the container, forcing the contents to escape from the second tube.
3. Because that tube is below the surface of the liquid, the liquid is forced out.

Because of the reservoir's capability to pump food-safe liquids, it's often used in drinkbots, also known as barbots. These are robots programmed to make cocktails by pumping liquor and mixers in precise amounts. The way it works is that the bottle (of rum, for instance) is the reservoir and gets pressurized with an air pump.

See "Mini Project: Make a Pressurized Reservoir" to learn how to use an old aquarium pump to make a plastic jug into a food-safe pressurized reservoir.

Peristaltic Pump

Another food-safe option is a peristaltic pump, which also never touches the liquid. Instead, it uses a motor to massage a tube, which forces the liquid to travel along the tube. This is similar to the way our gastrointestinal system works to, er, move food through. You most often see peristaltic pumps in breast milk pumps—the milk goes through food-safe tubing to stay pure for Junior.

As with pressurized reservoirs, you'll often see tinkerers using these pumps to make drinkbots. Peristaltic pumps are more expensive, but also much more controllable and precise in terms of stopping and starting the flow of liquid.

The pump pictured in Figure 7.4 is available from Adafruit Industries (P/N 1150). It's essentially a 5000 RPM, 12-volt DC motor with a "clover"-shaped hub that squishes the tubing and forces the liquid to move through it.

FIGURE 7.4 A peristaltic pump squishes a rubber tube to move fluid through it. Credit: Adafruit Industries.

Mini Project: Make a Pressurized Reservoir

Let's build an Arduino-controlled, pressurized reservoir! You'll use an aquarium bubbler to pressurize a milk jug, as shown in Figure 7.5. This forces its contents to dispense through the exit tube. Voila, a pump!

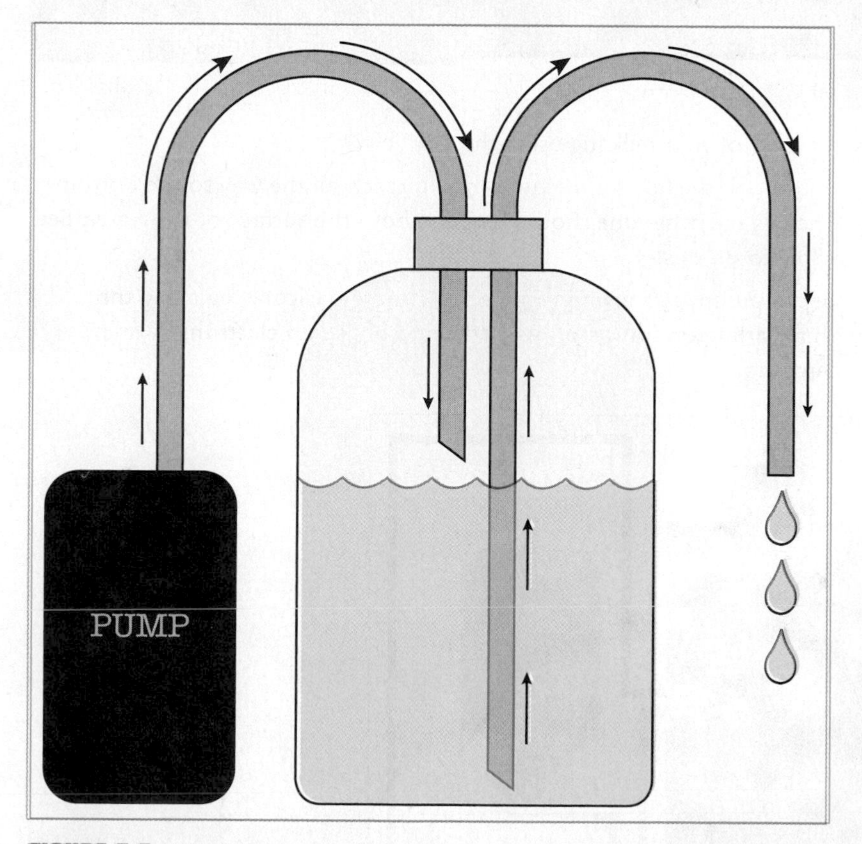

FIGURE 7.5 The pressurized reservoir displaces liquid by forcing air into the container.

PARTS LIST

- Arduino
- A milk jug—the classic gallon or half-gallon plastic jug works well
- A battery-powered aquarium pump (I used a Marina P/N 11134)
- Two lengths of tubing (I used Tygon B-44-3 beverage tubing with a 1/4" outer diameter)

- A TIP-120 Darlington transistor (Adafruit P/N 976; see Chapter 13, "Controlling Motors," to learn more about this useful component)
- 2.2K resistor
- 1N4001 diode (Adafruit P/N 755)
- Some wire (I recommend Adafruit P/N 1311)
- Drill and a 1/4" bit

Instructions

1. Drill two holes in the lid of your milk jug using the 1/4" bit.
2. Thread the tubing through the holes. One tube should reach all the way to the bottom of the milk jug, whereas the other one should remain above the surface of the water. See Figure 7.5 to see how to do this.
3. Wire up the aquarium pump, as shown in Figure 7.6. You're basically replacing the rocker switch with a Darlington transistor, which is kind of like an electronic switch activated by the Arduino.

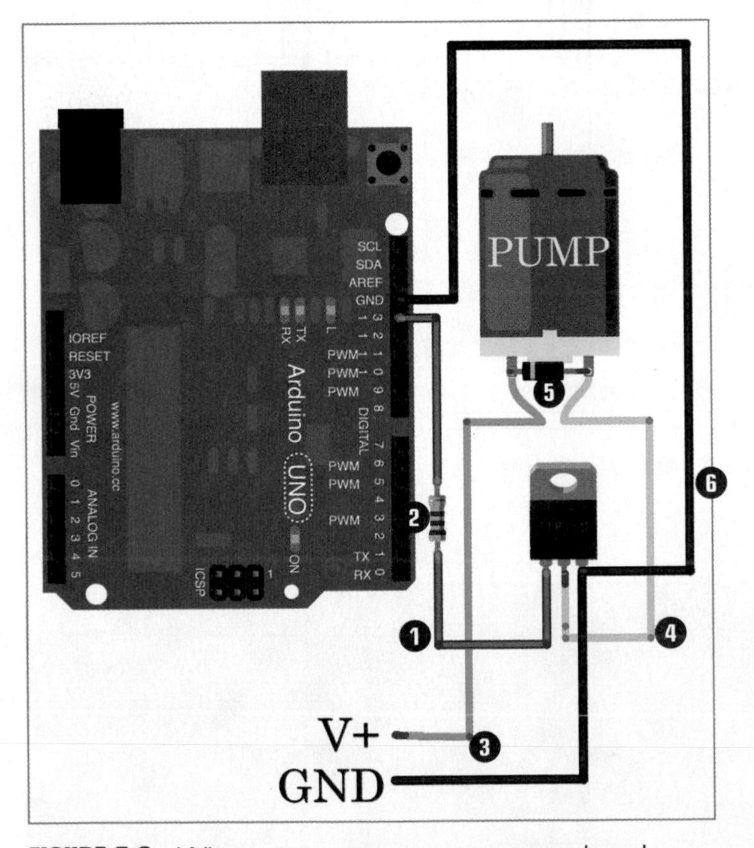

FIGURE 7.6 Wire up your pump as you see here!

1. Jumper wire
2. Resistor
3. Power source
4. Middle lead of Darlington goes to negative terminal on pump
5. Diode is soldered between + and - leads
6. Right lead of Darlington goes to the negative terminal of power supply and one of the Arduino's GND pins

a. Connect a jumper from Pin 13 of the Arduino to the leftmost lead of the Darlington transistor, with the resistor in between. This is the blue wire in Figure 7.6.

b. Connect the power source to the positive terminal of the pump. This is the green wire shown in Figure 7.6.

c. Connect the middle lead of the Darlington to the negative terminal of the pump; see the yellow wire in Figure 7.6.

d. Solder the diode between the positive and negative terminals of the pump leads; this helps prevent feedback from the motor frying your Arduino! The diode is polarized, so it must be attached correctly; Figure 7.6 shows how to do it.

e. Connect the right-hand lead of the Darlington to both the negative terminal of the power supply and one of the Arduino's GND pins. This is the black wire in Figure 7.6.

Pressurized Reservoir Code

I'm going to skip code because all you need to activate the pump is a single ping from your Arduino. I suggest using the Blink sketch from the Arduino's example code (see Chapter 5, "Programming Arduino," for more info) and changing the delays from 1,000 to 10,000. This makes the pump activate for 10 seconds, and then deactivate for the same amount of time, repeating until you unplug the board.

LEGO PERISTALTIC PUMP

Miguel Valenzuela (pancakebot.com) built a pancake-making robot out of LEGO bricks. It's pretty awesome and can "print" out letters and geometric shapes in pancake batter. Quite rightly, Miguel decided he needed a syrup-dispensing robot to accompany it, so he built a peristaltic pump out of LEGO bricks (see Figure 7.7) that sits next to PancakeBot, ready to squirt maple syrup onto your "printed" pancakes.

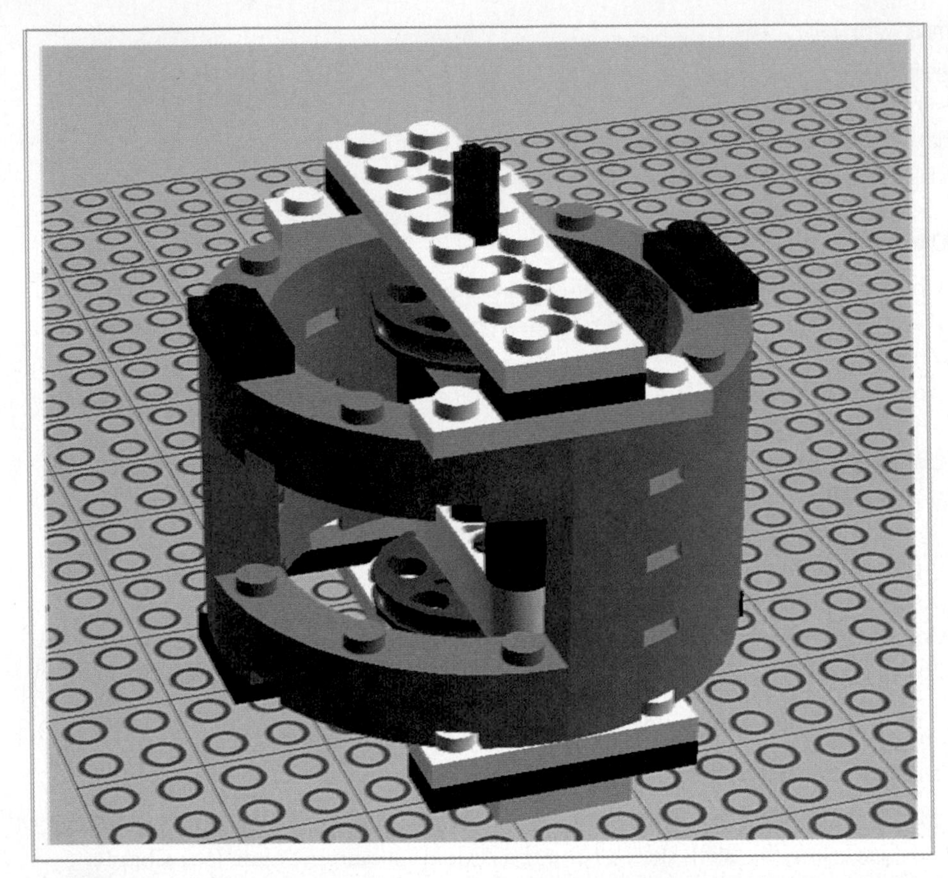

FIGURE 7.7 Miguel Valenzuela's syrup dispenser is actually a peristaltic pump. Credit: Miguel Valenzuela.

CAUTION

Food Safety

The important thing to remember in handling liquids is that if you're going to drink the liquid, it should only touch "food safe" surfaces. But what does that mean, exactly?

Regulations on food safety vary from nation to nation, but typically a food-safe surface must be free of contaminants that could hurt a person who eats or drinks off of the surface, and that through normal use, the surface won't decay in a way that contaminants could build up.

Most scientific and industrial surfaces that are food-safe identify themselves as such; if not, assume that it is NOT food safe.

Project: Plant-Watering Robot

For this chapter's project, you set up a robot that waters a plant on a set schedule, freeing you to make more labor-saving robots! The robot uses a solenoid valve (described earlier in this chapter) connected to a water supply (see Figure 7.8). It's controlled by an Arduino, which checks every minute to see whether it's time to dispense water. Finally, you'll build a nifty LEGO enclosure to protect the Arduino from sprays of water!

FIGURE 7.8 Water that hungry plant with this convenient plant-watering robot.

PARTS LIST

You'll need the following parts to build your plant-watering robot:

- Arduino Uno
- Solenoid valve (I used one from Adafruit, P/N 997, that is threaded for half-inch PVC fittings)
- TIP-120 Darlington transistor (Adafruit P/N 976)
- 2.2K resistor
- 9V battery
- 9V battery connector (Jameco P/N 109154)
- 1N4001 diode (Adafruit P/N 755)
- Some wire (I recommend Adafruit P/N 1311)
- 1 PVC elbow joint (The Home Depot [THD] P/N 406-005HC)
- 1 PVC elbow joint with threads (THD P/N 410-005HC)
- 1 PVC T-joint (THD P/N 401-005HC)
- 1/2" PVC tubing: I used the following lengths: 14", 21", and three lengths of 2.5" each (THD P/N 136293)
- PVC cement (THD P/N 308213)
- 2 threaded to regular adapters (THD P/N 435-005HC)
- A PVC end cap (THD P/N 448-005HC)
- PVC garden hose adapter (THD P/N 795399)
- A 3/4"-diameter dowel, sharpened on one end

PVC

A lot of makers swear by PVC (see Figure 7.9). You know, those white (usually) plastic pipes used for plumbing around the house. PVC is great for moving liquids around, but also finds a lot of use simply as a building material—it glues well and can be bent with heat and permanently glued.

PVC is often used for potato cannons and other "maker-y" projects involving moving substances such as air and water through the pipe, but what if you just want to make a chair out of your PVC? FORMUFIT is a company that offers a vast assortment of customized PVC connectors simply for making furniture. FORMUFIT's parts not only come in configurations you would never need for plumbing, but they're also less industrial looking. Most PVC you buy has obnoxious bar codes and other marks on the plastic and is very stark, the assumption being that it will be hidden under your sink, not looking good in your den. FORMUFIT's PVC, by contrast, is considerably more attractive, doesn't have writing on it, and comes in black and white.

FIGURE 7.9 Polyvinyl chloride (PVC) is both a common household building material as well as a handy maker's tool!

If you don't want to mail order your PVC, you can almost always find what you need (other than the specialized parts mentioned here) in your friendly local hardware store.

Instructions

Let's begin building the plant watering robot by starting with the PVC piping!

1. Grab the T-shaped PVC connector and connect the three 2.5" lengths of PVC, as shown in Figure 7.10. Secure the parts with PVC cement. The T should be arranged so one leg is pointing down, one pointing up, and one pointing backward.

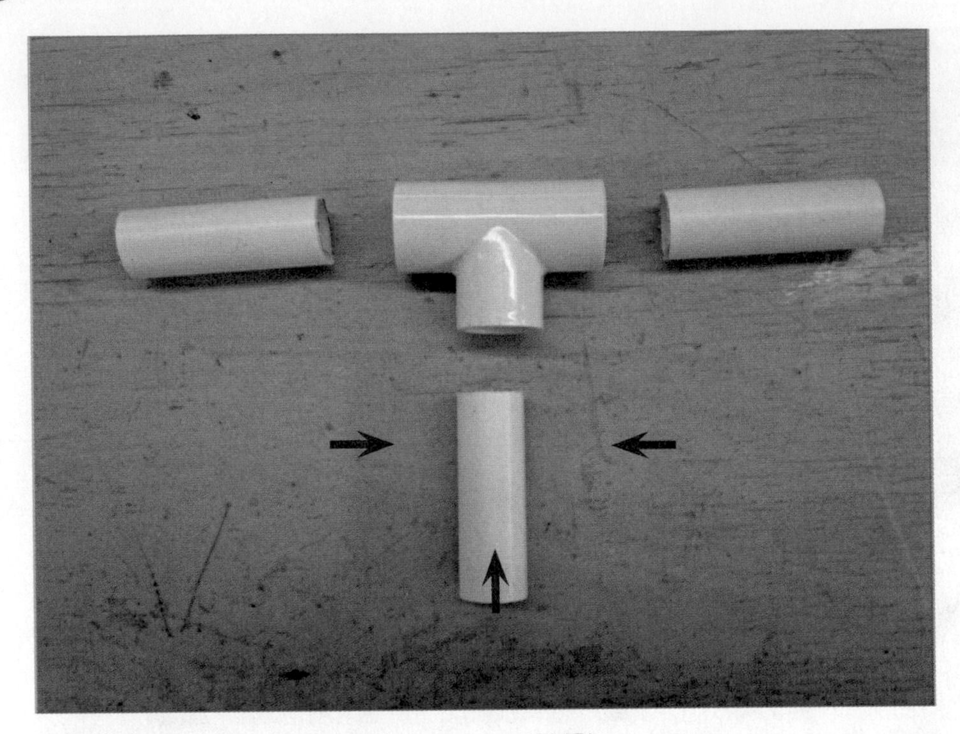

FIGURE 7.10 Start by cementing some PVC!

2. Add the dowel to the bottom-facing pipe of the T. The 3/4" dowel slides firmly into the inside of the 1/2" PVC pipe, and can be permanently connected with a smear of PVC cement. I'm not sure why a 3/4" dowel is smaller than a 1/2" pipe. Math, you're crazy sometimes!
3. You'll want to sharpen the other end of the dowel—I used a disk sander—so it'll stick into the ground and keep the PVC upright. The assembly should look like Figure 7.11.

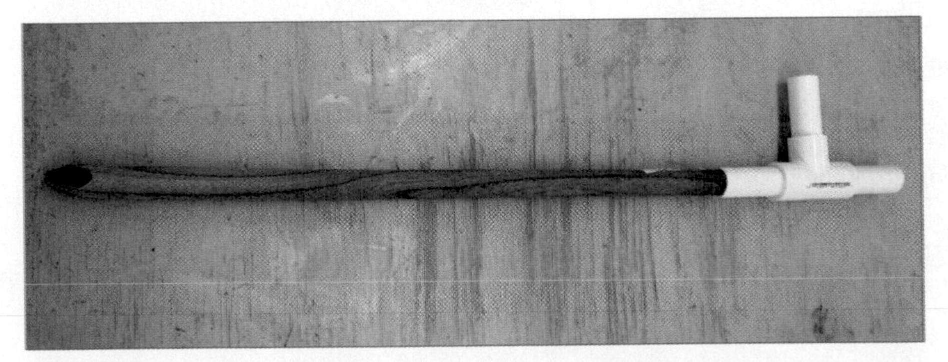

FIGURE 7.11 Cement in the dowel; this is the stake that keeps the robot anchored in the ground.

4. Connect the garden hose adapter to the arm of the T pointing backward, securing it with PVC cement. You can see the adapter in Figure 7.11.
5. Now work on the valve: Screw the two threaded adapters onto the ends of the valve, as shown in Figure 7.12. Note that the PVC threads aren't really intended for repeated opening and closing: After you tighten the threads, unscrewing the adapters will be very hard, so be absolutely certain you're ready to commit!

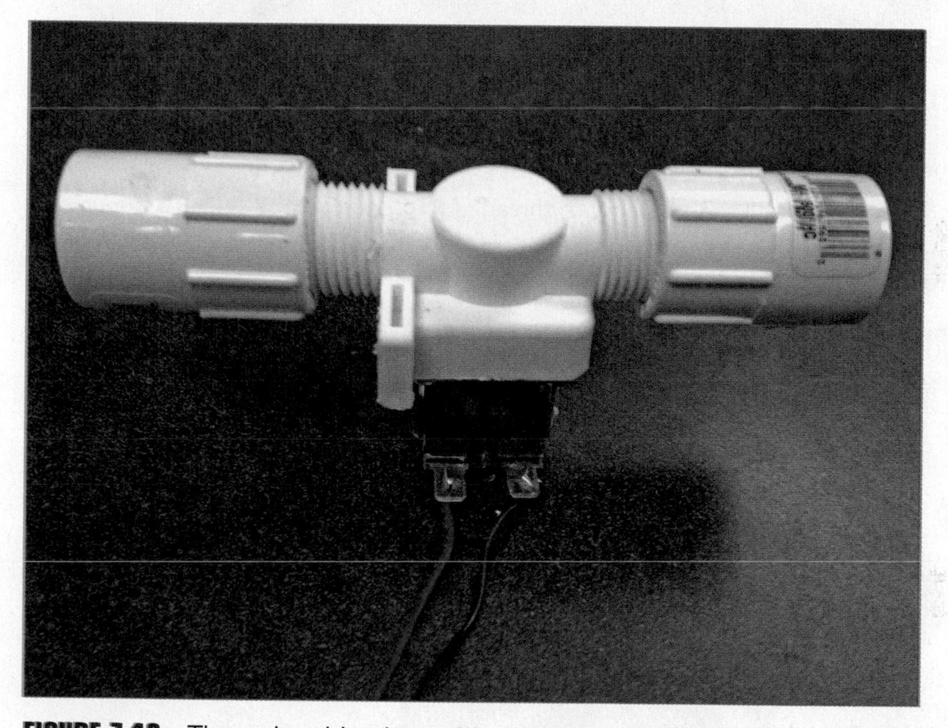

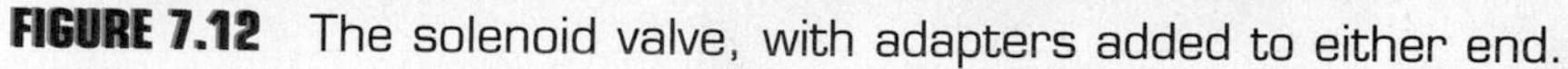

FIGURE 7.12 The solenoid valve, with adapters added to either end.

6. When you have the adapters firmly seated, cement the T assembly to the lower end, and the sprayer assembly to the upper end. But which is which? The valve is unidirectional, so you have to make sure the inlet is pointing down; look for a small arrow on the valve and make sure it's pointing up.
7. Take the 14" length of PVC tubing and cement one elbow joint to one end, and the other elbow to the other end. Figure 7.13 shows you how it should look.

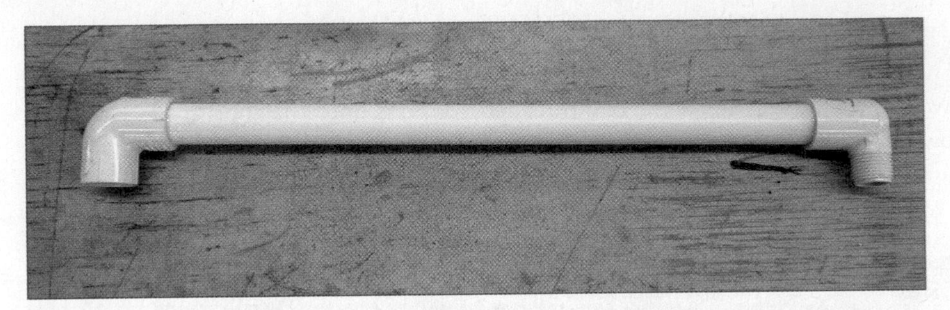

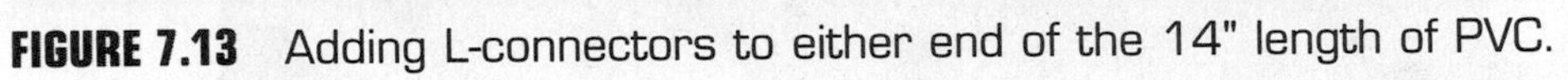

FIGURE 7.13 Adding L-connectors to either end of the 14" length of PVC.

8. Add the 21" length of PVC and cement it to the smooth (not threaded) elbow joint.
9. Now create the nozzle. I took a standard threaded PVC end cap and drilled a bunch of holes in it (see Figure 7.14). The size of the holes is dictated by how much water you want to come out!

FIGURE 7.14 Drill holes in an end cap to make a sprinkler head!

10. Attach the nozzle. Note that in Figure 7.15, I used a small extender module (THD P/N 434-005HC) but this is by no means necessary.

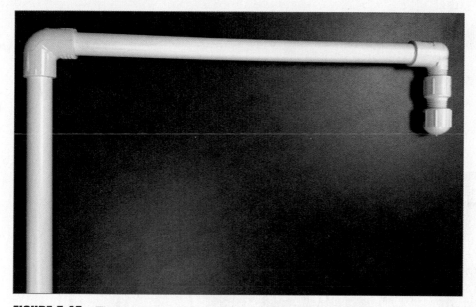

FIGURE 7.15 The top of the robot takes shape!

Plant-Watering Robot Electronics

Now it's time to wire up the electronics, which you do pretty much the same way that you wired up the pressurized reservoir described earlier in the chapter. See Figure 7.16.

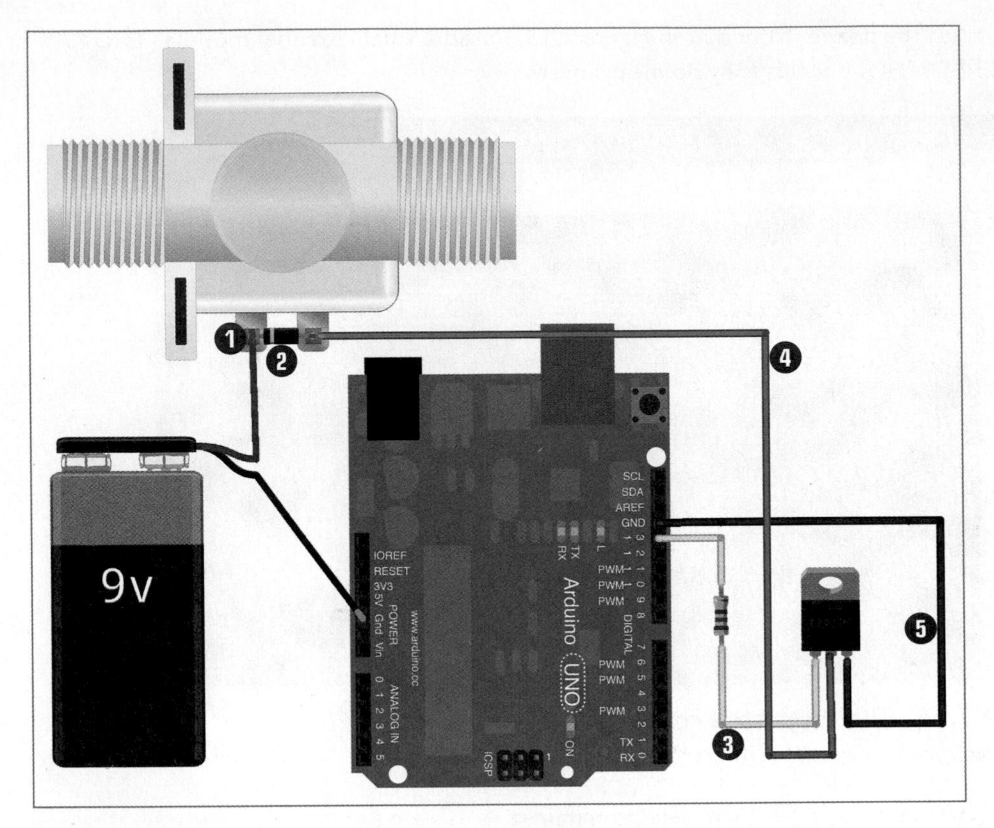

FIGURE 7.16 This diagram shows you how to wire up the valve.

1. Connect the + battery lead to the + solenoid valve lead
2. Connect + and - valve leads with diode
3. Arduino pin 13 connects to left lead on transistor
4. Connect - terminal to center pin on transistor
5. Arduino GND connects to right transistor pin

Here's what you do:

1. Connect the positive lead of the 9V battery pack to the positive lead of the solenoid valve. This is the red wire in Figure 7.16.
2. Connect the positive and negative leads of the valve with a 1N4001 diode, which helps protect the other electronics from stray voltage from the motor. The stripe on the component should be pointing toward the positive lead.
3. Connect pin 13 of the Arduino to the leftmost lead of the Darlington transistor, with a 2.2K resistor in between. This is shown as a yellow wire in Figure 7.16.
4. Connect the negative terminal of the valve to the center pin of the Darlington transistor. This is the blue wire in Figure 7.16.

5. Connect the rightmost pin of the Darlington transistor to GND (the black wire in Figure 7.16).

Plant-Watering Robot Enclosure

Now you can build a LEGO enclosure that will house the Arduino and other electronics and protect them from the elements. Figure 7.17 shows the enclosure design.

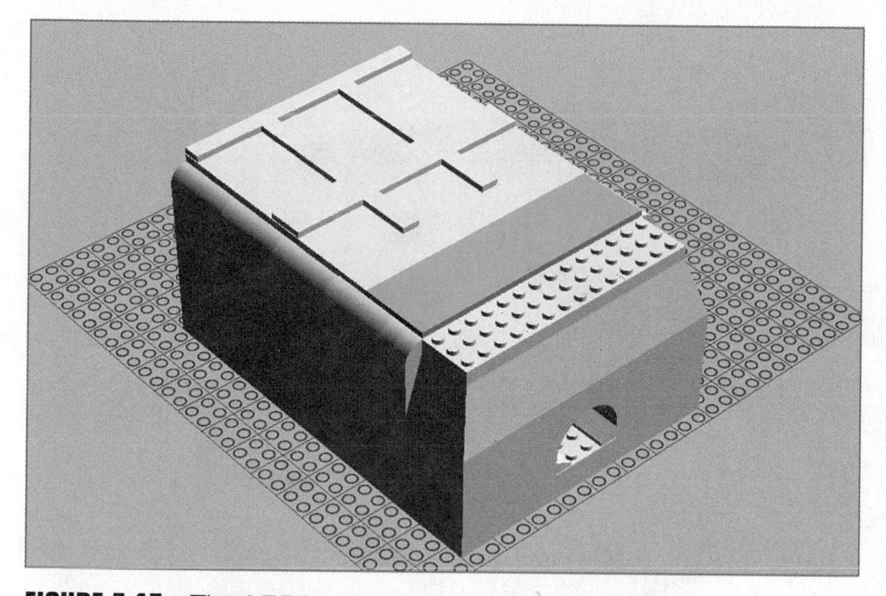

FIGURE 7.17 The LEGO enclosure, as rendered by a computer.

You can find instructions on how to build the enclosure shown in Figure 7.17 in the form of a LEGO Digital Designer (LDD) file at https://github.com/n1/Arduino-For-Beginners.

You can download LDD for free from ldd.lego.com, and it runs on any reasonably modern PC or Mac—sorry Linux heads! The file consists of a 3D CAD drawing of the model, and you can tell LDD to create step-by-steps for you right out of the program. It's slick! There are actually two files, one for the top of the box and one for the bottom. The instructions show you how to build the enclosure in two parts so it can encircle the PVC.

When building the enclosure, you might want to consider whether you want to—brace yourself—GLUE the bricks together. LEGO fanatics abhor gluing their bricks, but if you're serious about keeping the interior dry, you just might have to. Plus, it prevents the enclosure from breaking apart randomly!

Adding the Electronics

You should consider how to attach the Arduino, battery, and associated circuitry to the enclosure. One option might be to use zip ties to secure the board to one of LEGO's Technic bricks (see Figure 7.18). These look like regular LEGO bricks but have holes in them, making them great for zip ties! You then simply connect the bricks to the inside of the enclosure as you would do any LEGO brick.

FIGURE 7.18 Secure the Arduino to the enclosure.

If you don't have any Technic bricks, another option might be to use an adhesive to attach the circuit boards. Maybe double-sided tape? In Chapter 2, "Breadboarding," I used super-cool putty called Sugru (sugru.com) to attach an especially tricky component to the enclosure. This is not a bad option in this project. Another use of the putty could be to seal the gaps around the PVC pipe to keep out moisture, as shown in Figure 7.19.

FIGURE 7.19 Smear in some Sugru to help the enclosure resist water.

After you've connected the two halves of the LEGO enclosure, goop some Sugru in the space around the PVC. After it sets (24 hours), it will not only keep out moisture, but will also inhibit the LEGO box from rotating.

NOTE

Why Do It That Way?

The keen-witted among you are likely saying, "Why not simply drill a hole in another material closer to the diameter of the PVC?" True, I could certainly do that, but I wanted to use LEGO for this project.

If you've followed along with these instructions, the end result should look like Figure 7.20. Congrats, and get to watering!

FIGURE 7.20 You're finished!

PROTOTYPING WITH LEGO

A lot of serious engineers (seriously) use LEGO bricks as part of the prototyping process. It's quick, has no learning curve, and it's often simply lying around waiting to be used. Why spend money on something slower and more expensive?

As you saw from this project, you can easily make LEGO boxes. However, you can build extremely complicated robots using LEGO Mindstorms robotics set. Want to build a robot? Consider nailing down the design with LEGO Digital Designer first (see Figure 7.21).

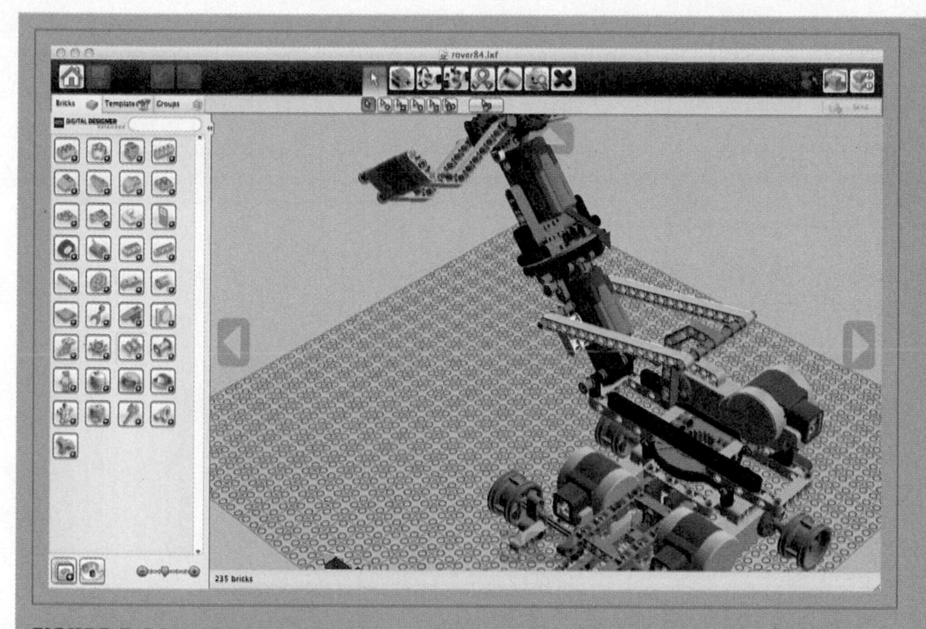

FIGURE 7.21 Want to build a robot virtually? LEGO Digital Designer lets you build online!

Some people build with LEGO Digital Designer with every intention of replacing it with a "real" enclosure. I often find the box I had in mind for a project ends up being the wrong size. With a LEGO enclosure, you'll already know the perfect dimensions before you start building the final version of the project!

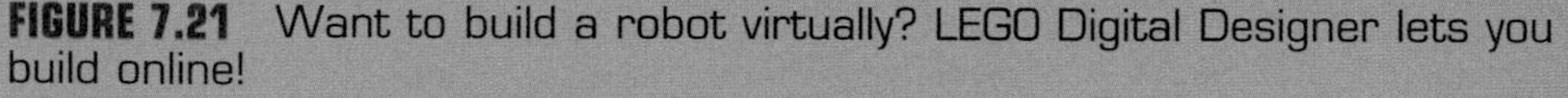

Plant-Watering Robot Code

The plant-watering robot has simple code. Just as in the sample project earlier in the chapter, this project uses a modified Blink sketch, which simply turns on pin 13 for a period of time, and then deactivates it for another period of time. Because all you need is one pin to trigger the valve, it's not a complicated program. The most interesting part (for me) is the timing. I created variables that can be set by you to control how often the water dispenses, and rather than using the rather unwieldy milliseconds the Arduino looks for, these variables use hours and minutes.

NOTE

Code Available for Download

You don't have to enter all of this code by hand. Simply go to https://github.com/n1/Arduino-For-Beginners to download the free code.

To learn more about how an Arduino keeps track of minutes and seconds, see Chapter 11, "Measuring Time."

```
int valve = 13; // renames Pin 13 "valve"

int offhours = 0; // how many hours before the water dispenses?
int offmins = 1; // how many minutes before the water dispenses?
int spray = 10; // number of seconds the water sprays

void setup() {
  pinMode(valve, OUTPUT);  // designates the valve pin as "output"
  Serial.begin(115200);
}

void loop() {
  int wait = (offmins * 60000) + (offhours * 3600000); // computes milliseconds

  digitalWrite(valve, HIGH);
  delay(spray * 1000); // water stays on this number of milliseconds
  Serial.println(offmins * 60000); // I used this when debugging
  digitalWrite(valve, LOW);
  delay(offmins * 60000); // water stays off this number of milliseconds
}
```

The Next Chapter

Let's talk about tools. Chapter 8, "Tool Bin," explores a lot of the tools I used to prototype and create the projects in this book, as well as some related gear that you're likely to encounter in a well-equipped workshop.

8

Tool Bin

You must have noticed thus far in the book that you need a lot of tools (see Figure 8.1) to build the various projects. In this chapter, you'll explore some of the equipment that conceivably you might need, beginning with my take on the ultimate maker's toolkit. After that, I detail multimeters, those invaluable measuring devices that electronics hackers swear by. You'll then learn about various tools you need to work in wood, plastic, and metal. I know this all costs a lot of money, so I provide suggestions on how to get access to tools without a huge investment. Finally, because this is primarily an electronics and Arduino book, I provide several tips on hobbyist electronics such as mastering multimeters, harvesting electronics from scrapped devices, and identifying mysterious parts by their markings.

FIGURE 8.1 It's surprising how much stuff an ordinary toolbox can hold.

Maker's Ultimate Toolbox

You need a bunch tools, but what exactly? The equipment you need in your main toolbox (see Figure 8.2) will vary—toolboxes are as unique as the people who use them, and ultimately only you can decide what you need. That said, here are some ideas for what you might want to consider including.

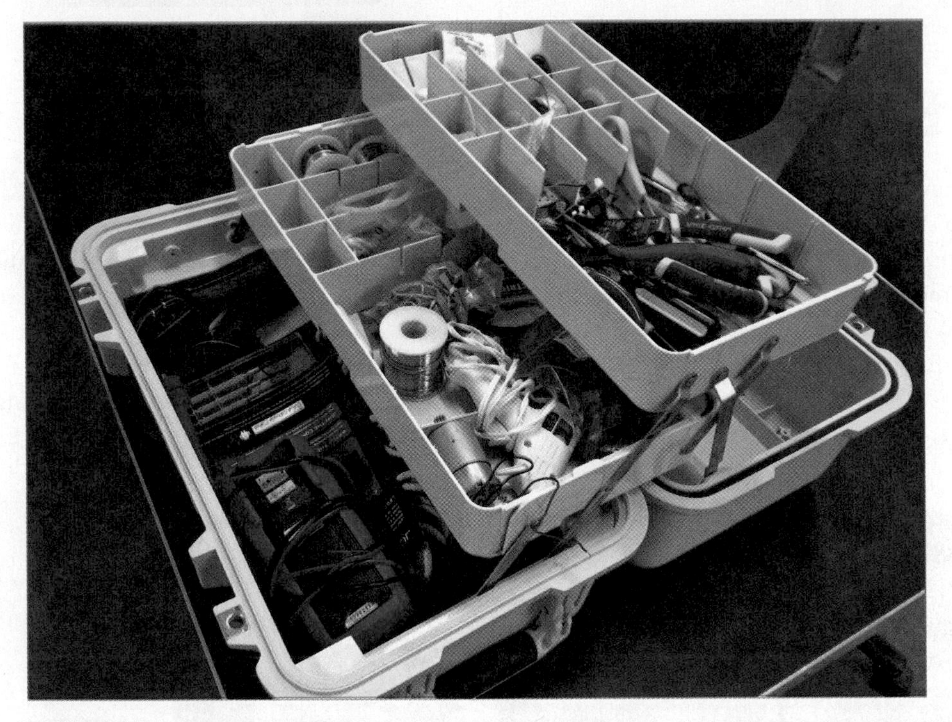

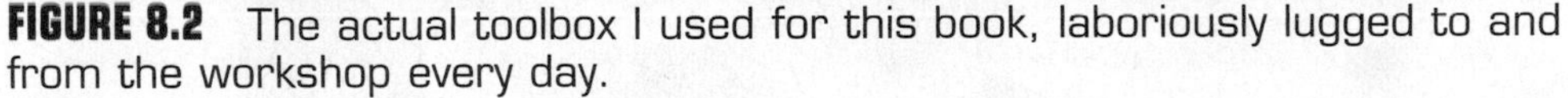

FIGURE 8.2 The actual toolbox I used for this book, laboriously lugged to and from the workshop every day.

Basic Multimeter

The number-one diagnostic tool used by electronics tinkerers is a multimeter. Meters are sophisticated measuring devices used to take voltage readings, measure resistance, test connectivity, and so on. In the "Electronics Tools and Techniques" section later on this chapter, you'll learn all about these handy devices. In the meantime, I suggest including a cheap meter in your toolbox. I use the DT-830B (shown in Figure 8.3) for my "on-the-go meter" and you can buy it at any number of online stores.

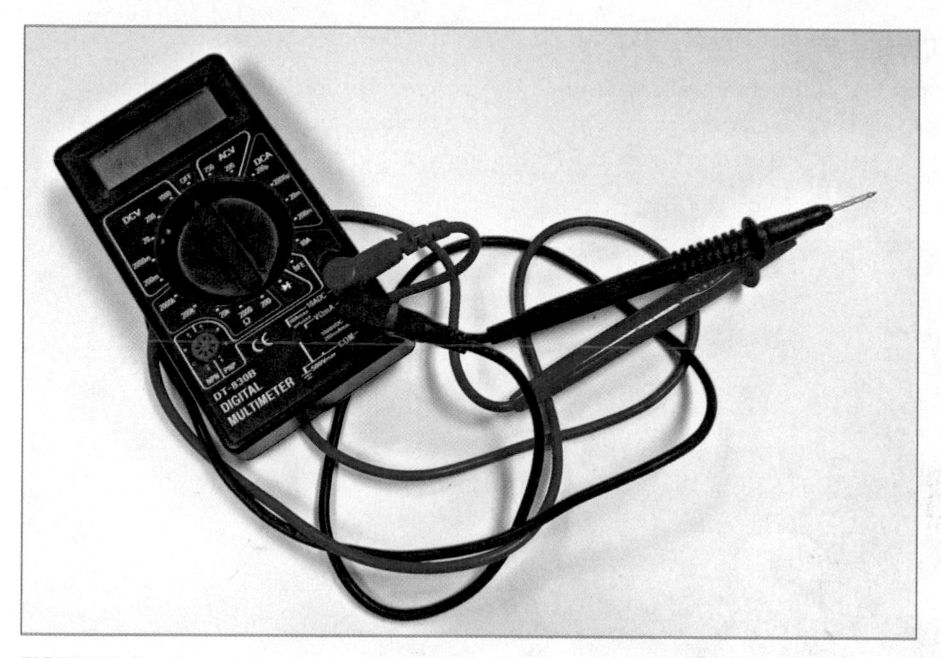

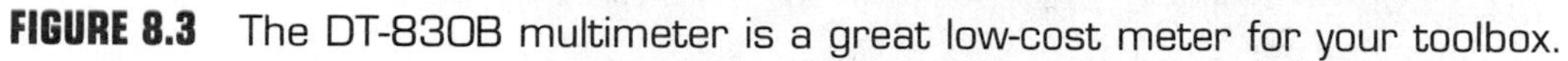
FIGURE 8.3 The DT-830B multimeter is a great low-cost meter for your toolbox.

Multitool

The handy multitool device consists of several tools in one package, hence the name. Multitools usually have blades, screwdrivers, files, pliers, wire cutters, and other necessary tools. My own tool, the SOG Knives B61 (see Figure 8.4), gets used pretty much every day.

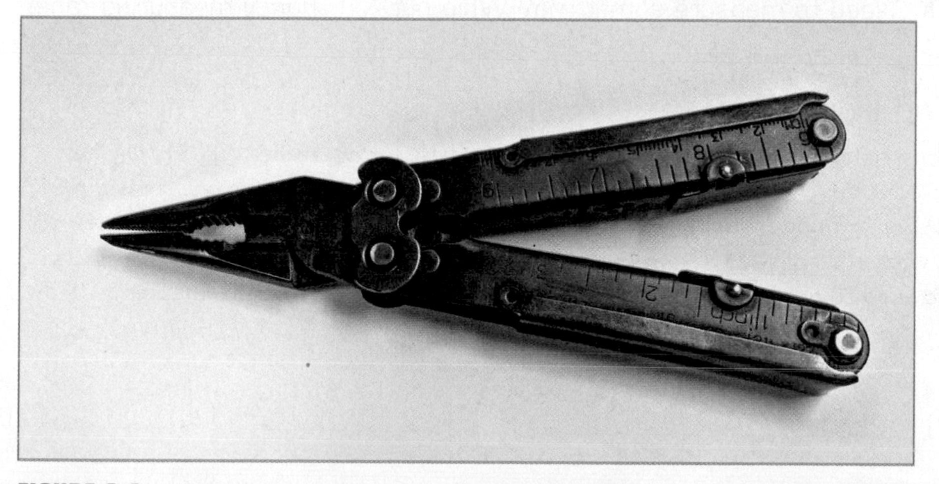

FIGURE 8.4 Multitools combine several tools into one tool, giving you a lot of options for tackling a project.

Measuring Tape

You might feel like a carpenter walking around with a tape measure, but it will definitely come in handy. I keep one at the workshop, one in my toolbox, and one at home. I like the Stanley tape measure pictured in Figure 8.5 (the Home Depot P/N 33-425D)—it's as robust as I like but only $10!

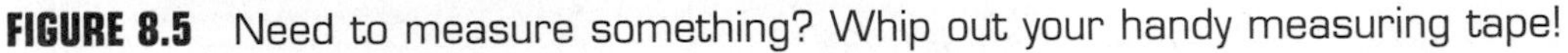
FIGURE 8.5 Need to measure something? Whip out your handy measuring tape!

Soldering Iron

I like a robust soldering iron, and have nothing but praise for my Weller WES51 (DigiKey P/N WES51-120V-ND) but it's kind of huge. A pen-style iron (such as the one shown in Figure 8.6) with no power supply, stand, or sponge fits nicely in your toolbox for those occasions when the full-sized iron is impractical; Adafruit offers a good model (P/N 180). You're also likely to need additional soldering equipment, which I detail in Chapter 3, "How to Solder."

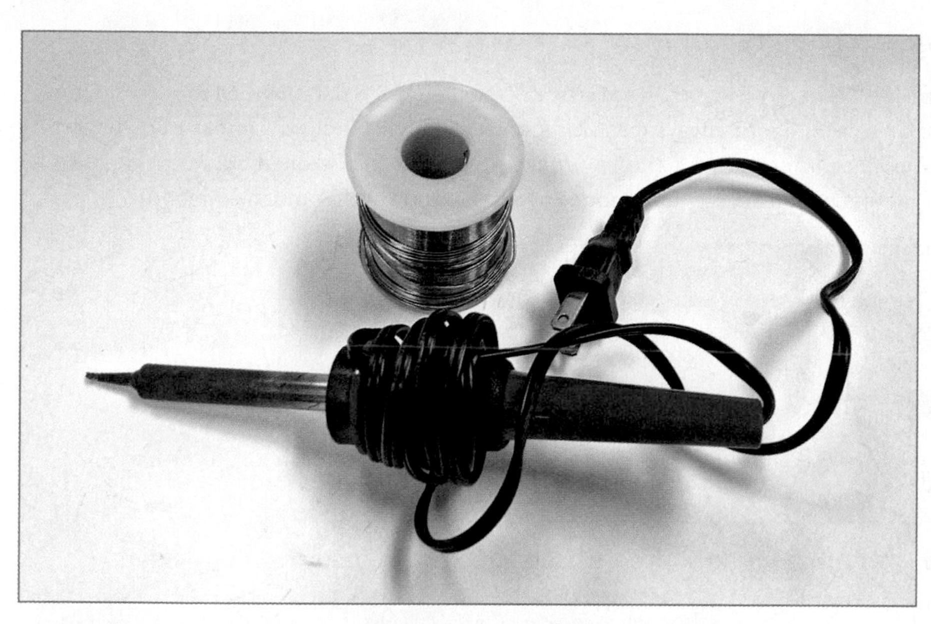

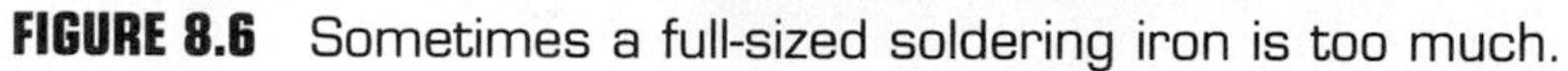

FIGURE 8.6 Sometimes a full-sized soldering iron is too much.

Digital Caliper

Measuring tapes are well and good, but sometimes you need to measure something precisely, and that's where a caliper comes in. The one I use has a digital readout (shown in Figure 8.7) that displays precisely the distance between the two prongs. It's a Neiko 01407A and costs around $17, a steal for a tool this useful!

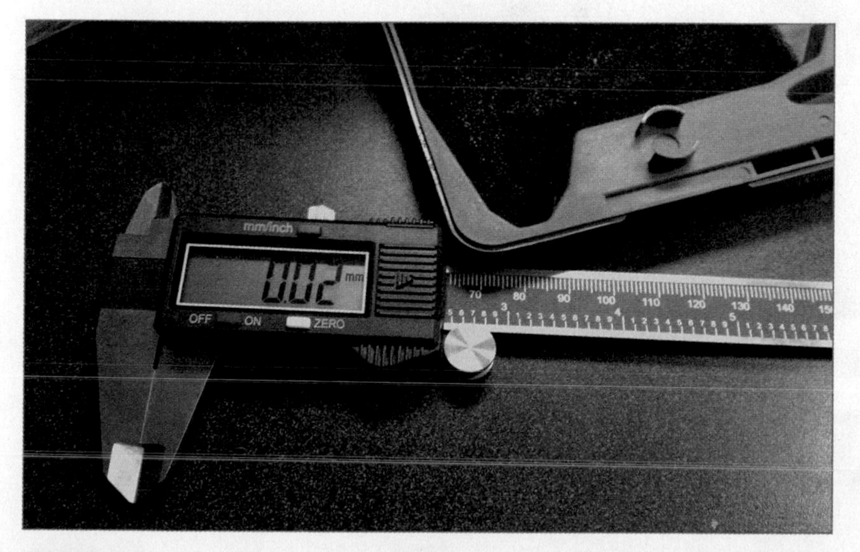

FIGURE 8.7 A digital caliper displays the measurement on a LCD screen—no guessing!

Scissors

They might seem rather pedestrian, but scissors (see Figure 8.8) are a great tool to keep in your toolbox. Yes, you could always cut with a knife or other blade, but most of us have extensive experience using scissors and can make precise and controlled cuts with them. A decent pair won't set you back much money and they don't weigh much—why not?

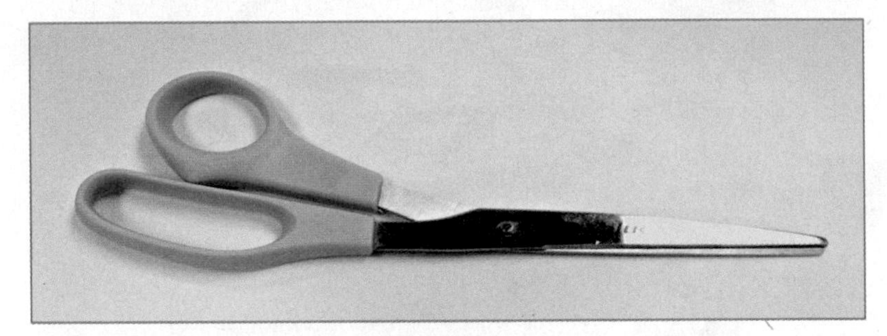

FIGURE 8.8 It's amazing how often scissors come in handy.

X-Acto Knives

Another good cutting tool are hobby knives, often called X-Acto knives (see Figure 8.9) because that company, now owned by Elmer's, has become synonymous with that kind of cutting implement. When you need something cut precisely with a very sharp knife, go with one of these tools.

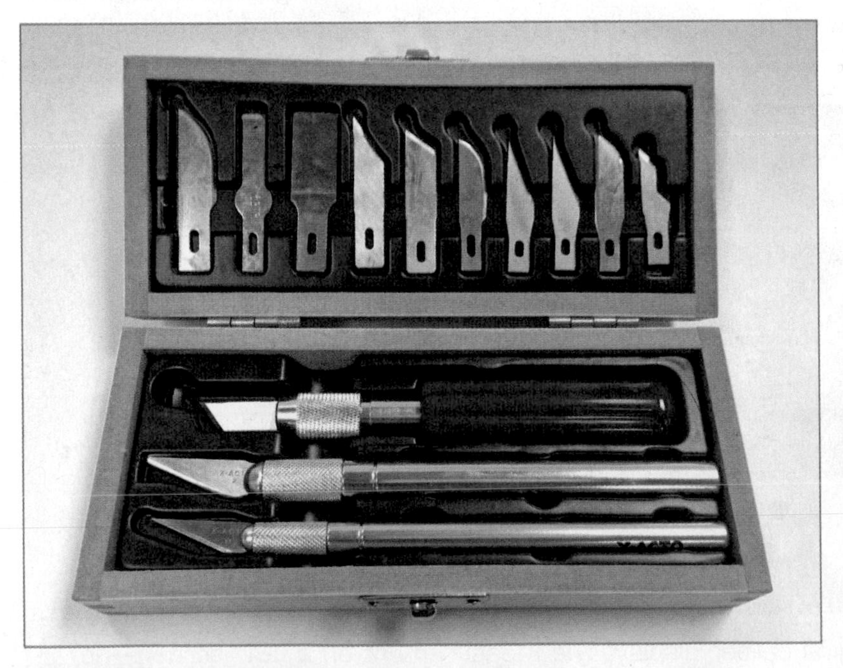

FIGURE 8.9 X-Acto knives are great for precision cutting.

Screwdrivers

I'm always reaching for a screwdriver, because there's always a screw to tighten or loosen. You'll definitely want a variety of drivers, like those shown in Figure 8.10. Especially be sure to get a set of smaller drivers, because maker projects often use small hardware. You'll also want a wide variety of formats such as Phillips, hex, Torx, and so on. The set pictured in Figure 8.10 (Fuller Tool P/N 135-0916) is a good starting place. It has 16 bits and several different tip styles, and it's about $10 on Amazon.

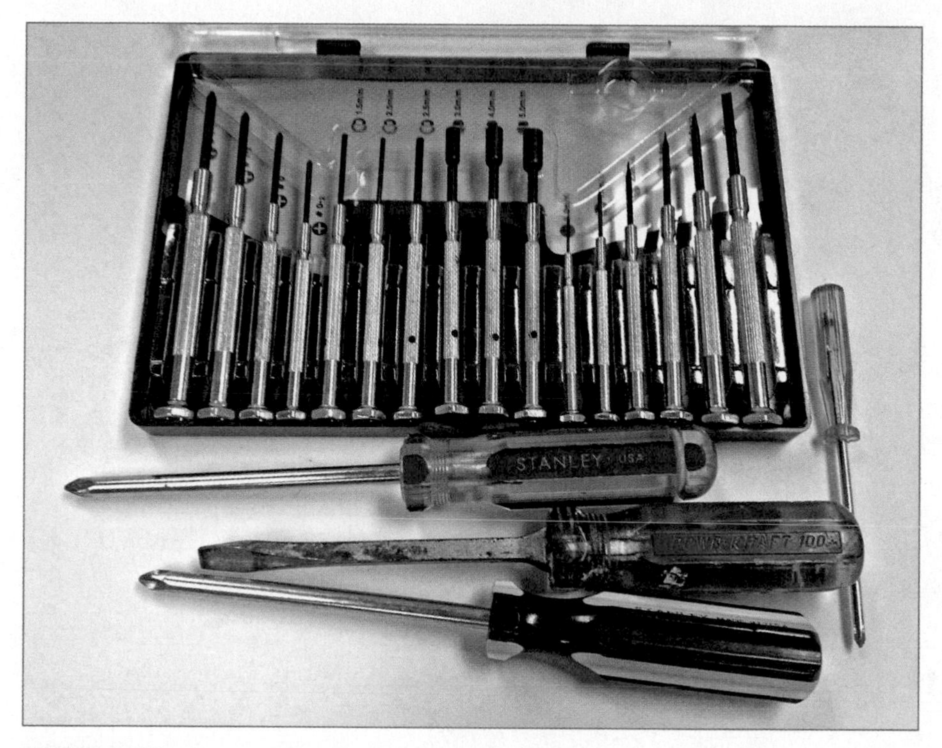

FIGURE 8.10 Screwdriver set.

Hardware

You also need screws, bolts, washers, and so on. I'm always reaching into my hardware jar (see Figure 8.11) for connectors, because it's easier and faster than running to the store. A peanut butter container makes a great container for storing nuts and bolts, and it fits into many toolboxes.

FIGURE 8.11 Miscellaneous hardware in a plastic jar.

Wire Strippers

Another tool that you'll barely put down is the combination wire cutter and stripper like the Vise-Grips in Figure 8.12. I use these things all the time and they're probably the first tool I grab on any particular day. You can get these for under $10 from Amazon and other online stores.

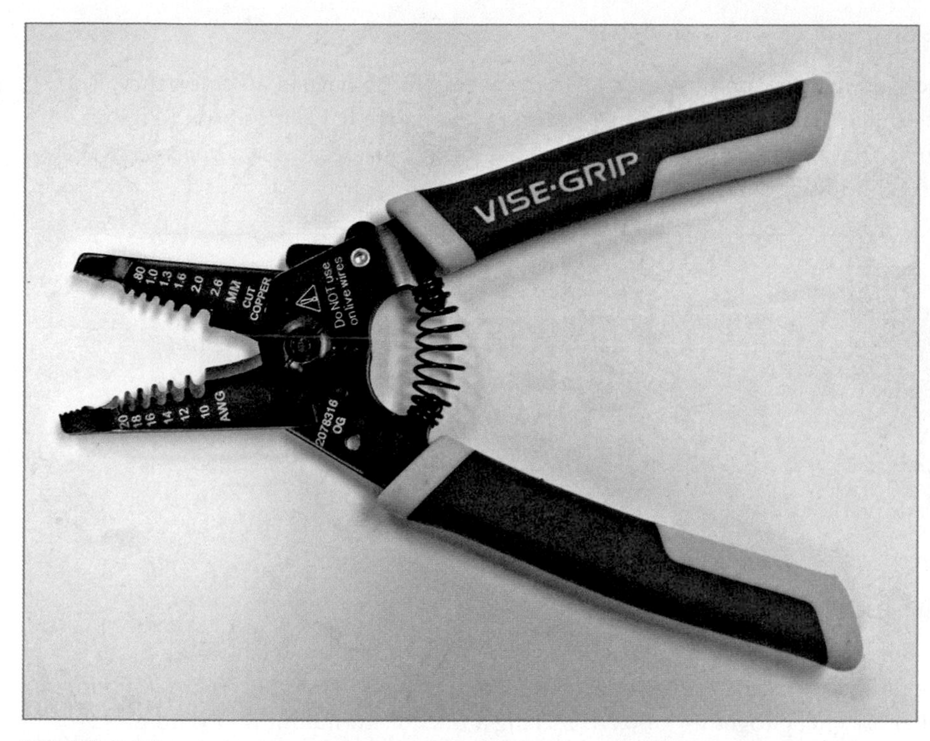

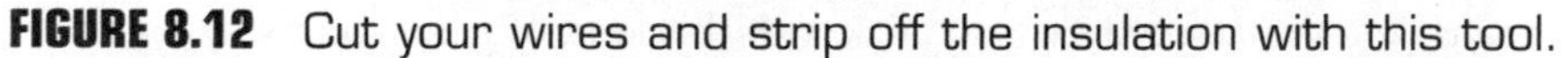

FIGURE 8.12 Cut your wires and strip off the insulation with this tool.

Super Glue

Super glue (often sold as Krazy Glue, see Figure 8.13) bonds pretty much anything to anything! This is one of those things you won't need most of the time, but when you do need it, you'll totally want a fresh tube in your toolbox. I say fresh because after you open a tube, the chances of it still being useable the next time you need it is not good.

FIGURE 8.13 Every so often you'll REALLY need super glue.

Mini Flashlight

Another tool you mostly won't use, but will be very grateful to have in an emergency, is a small pen flashlight, such as the Pelican 1920 shown in Figure 8.14. It's about $20, but you can find similar flashlights for much cheaper. I like this model because it's built tough, with a stainless steel body, and it has a handy pocket clip.

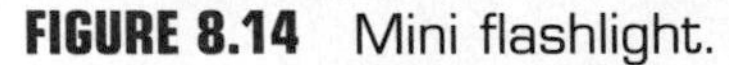

FIGURE 8.14 Mini flashlight.

Hot Glue Gun

When in doubt, just glue it! A hot glue gun, like the mini model shown in Figure 8.15, is possibly the ultimate maker's tool—at least for temporary fixes! Hot glue comes in handy a lot, whether for gluing together a box or tacking a difficult-to-mount part onto an enclosure. That said, know that anything you stick with this kind of glue won't stay stuck—hot glue is for temporary fixes only.

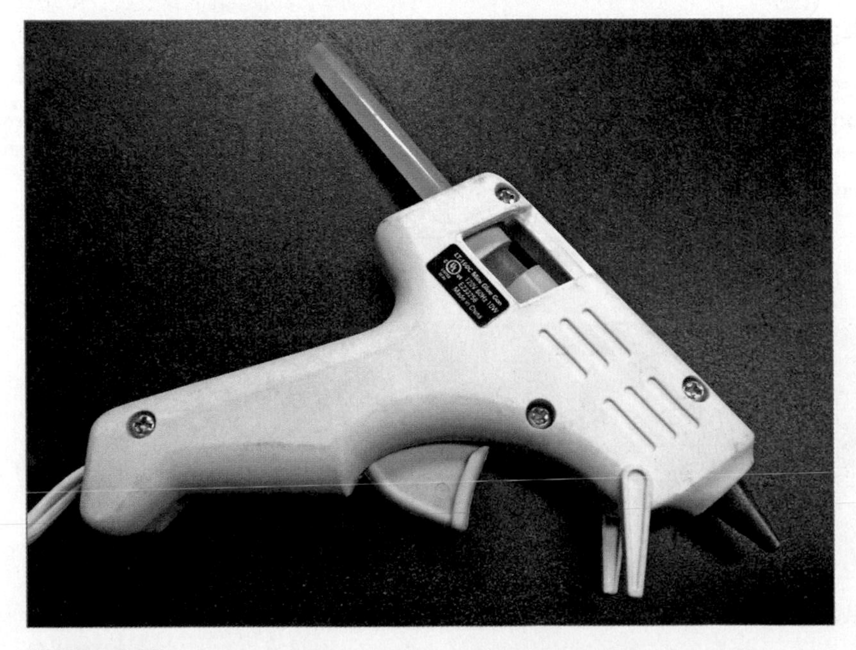

FIGURE 8.15 A hot glue gun, the maker's secret weapon!

Magnifying Glass

Maybe it's just me, because I recently started needing bifocals, but often times a magnifying glass (see Figure 8.16) comes in handy. Here's an example: Suppose you solder up a circuit board and it doesn't work. Being able to inspect the solder traces with a magnifying glass greatly speeds up the debugging process.

FIGURE 8.16 Magnifying glass.

Writing Supplies

You need a wide variety of writing utensils (see Figure 8.17) to mark materials for cutting or drilling, to jot notes, or best of all, for sketching out your projects before the build begins.

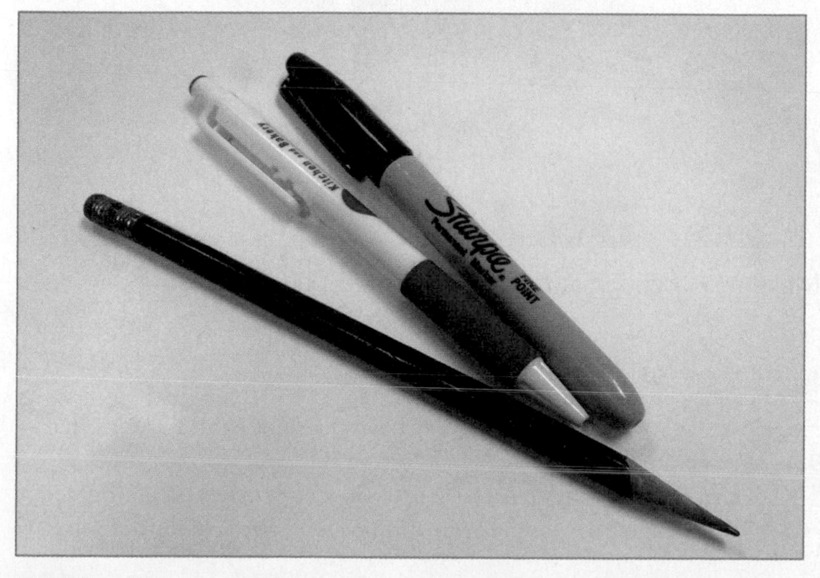

FIGURE 8.17 A variety of writing utensils.

Sketchbook

You'll also need a sketchbook in which to scribble your ideas. I like the Maker's Notebook (Maker Shed P/N 9780596519414; see Figure 8.18) because it has graph-lined pages, quick reference guides in the back, and comes with stickers so you can customize its appearance. The Maker's Notebook costs about $20, which might seem steep. I actually use ordinary composition books for most of my notes, and it's a no-frills but very inexpensive experience.

FIGURE 8.18 The Maker's Notebook is more than just a pad of paper.

Charging Cables

I don't know how much time I've wasted because I don't have the right cables, like those shown in Figure 8.19. You'll definitely want an Arduino-compatible wall wart (SparkFun P/N 298) as well as a standard USB cable (Adafruit P/N 62) in your toolkit. Also, don't neglect phone-charging cables. Keeping a spare in your toolbox will save you tons of hassle, trust me!

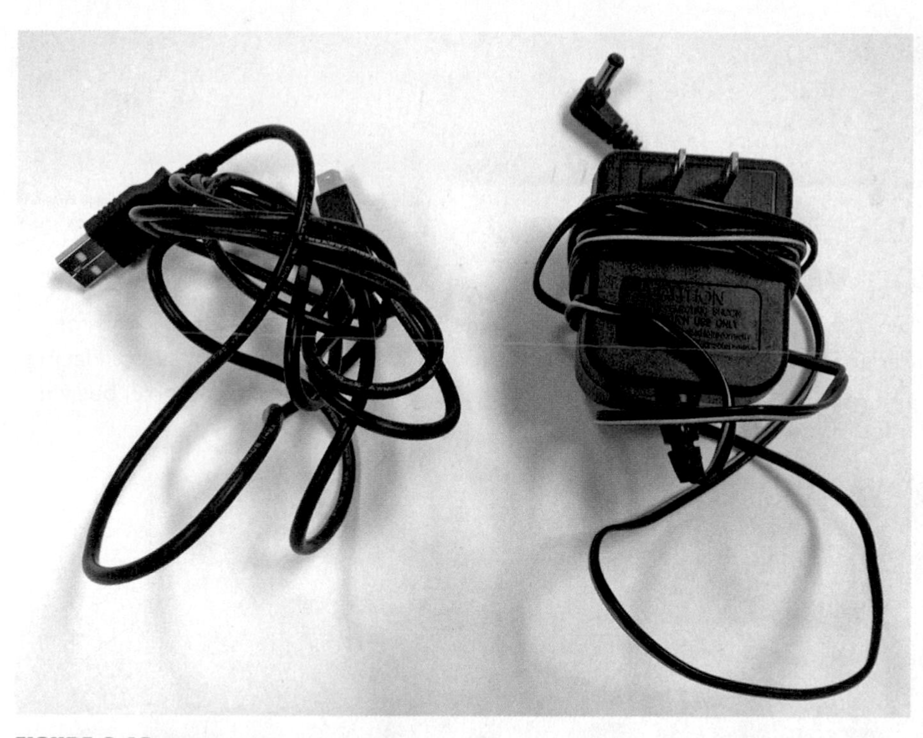

FIGURE 8.19 A USB cable and Arduino-compatible wall wart are must-haves in the maker's toolbox.

BASIC MAKER'S FIRST-AID KIT

Hopefully you'll never get seriously injured while working, but you should ensure you have a bare minimum of first-aid supplies on hand just for day-to-day cuts and scrapes. It's amazing how often minor injuries take place in a workshop, whether it's getting burned with a soldering iron to getting scraped by a saw blade. Here are some items to consider including in your first-aid kit:

- Adhesive bandages: The classic Band-Aid-style bandage, in a variety of sizes.
- Antibacterial ointment: Slather it on everything!
- Disinfecting wipes: Great for clearing the skin around a wound before treating the wound!
- Eyewash: These come in one-use bottles of sterile liquid, and if you get something in your eye, you can squirt it out.
- Hand sanitizer: Great for cleaning your hands before treating a wound. Not so great if the wound is on your hand!
- Hydrogen peroxide: An easy way to disinfect a big scrape, but, ow! It can kind of sting.

- Gauze: Comes both in squares and rolls of ribbon and is great for binding up bigger wounds.
- Medical tape: This is used for taping up gauze bandages.

Working with Wood

A lot of makers use wood in their builds for the same reason we've always used wood for building materials: because it's readily available, inexpensive, and easy to work with. Having access to a full wood shop (see Figure 8.20) makes making much easier, of course, but you can still do a lot of fun stuff at home.

FIGURE 8.20 Having access to a wood shop makes making easier!

The following sections discuss a number of tools that you might find in a well-equipped wood shop. This is the equipment I find myself using the most.

Laser Cutter

A laser cutter or laser etcher (see Figure 8.21) is a big machine with a precisely controlled laser that follows a path laid down in the software. If you want to build a box, all you have to do is design a box in a vector software program such as CorelDRAW or Adobe Illustrator, cut it out in the laser, and you've got yourself an enclosure! There really isn't a faster way to whip up a quick project box. The downside? Well, most people don't have access to a laser. Never fear, however; in the "Lasering and CNCing Services" section later on this chapter, I'll go over some ways of having someone else do the work for you.

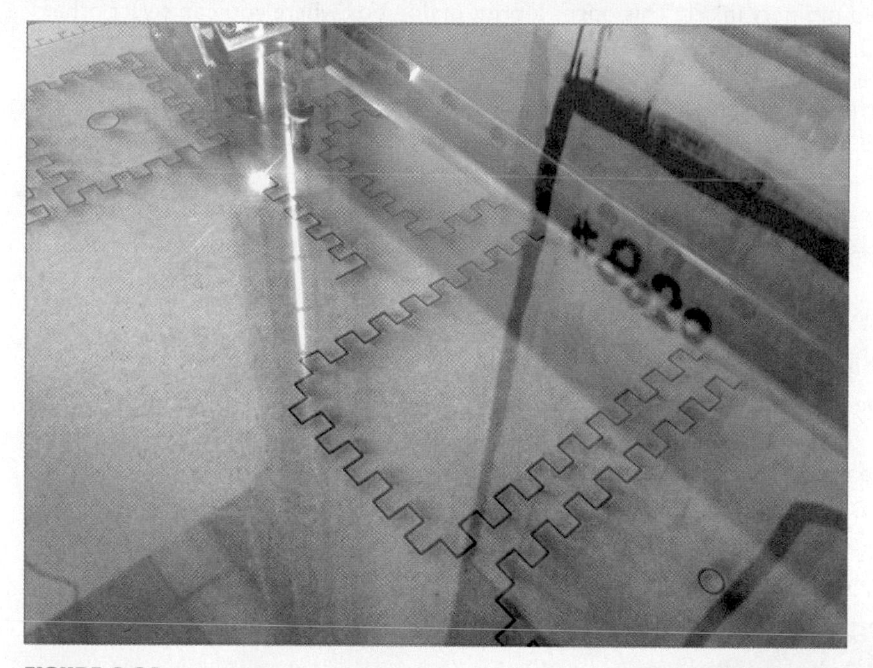

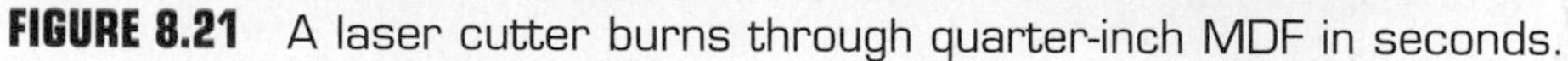

FIGURE 8.21 A laser cutter burns through quarter-inch MDF in seconds.

How to Use a Laser Cutter

Using a laser cutter can be extremely easy. I say "can" because they're all different. Every system has a propriety interface so creating a single guide on how to cut stuff with a laser is difficult. However, here are tips on using an Epilog, which is the most common brand in the U.S.:

- Prepare your vector file in Illustrator or CorelDRAW. Vectors are merely lines or paths, expressed as a series of curves. All lines to be cut should be hairlines (the skinniest width) and all shapes to be etched or engraved (rather than cut) should be raster images such as photos or logos. These will be burned into the wood but shouldn't go all the way through.
- Select your material and place it on the laser's bed. I had great luck with 1/4" MDF as well as composition board and acrylic. However, the glue in plywood diffuses the laser, inhibiting cutting and creating a huge amount of char.
- On the laser's accompanying PC, launch the vector design software (the one I used had CorelDRAW on it) and open the file you intend to cut. Go to Settings and make sure

you have the main three settings configured the way you like it: A) Speed, or how fast the laser moves around, B) Power, how strong the laser burns, and C) Frequency, which is how fast the laser pulses on and off.

- When you have your material on the bed, with the design ready to cut, click Print as you would with an ordinary inkjet. This opens a print dialog box where you can select other options. Click Print, and the vectors will be sent to the laser!

Using a laser cutter is actually very easy and chances are, the biggest problems you'll encounter will be using the wrong settings and charring or melting what you're trying to cut. There's a certain amount of experimentation involved! Part of the laser cutting experience consists of playing with your settings to get the right cut. Don't be dismayed if your material scorches, or the laser doesn't make it all the way through. Simply tweak your settings and try again.

Rotary Tool

In contrast to the laser cutter, a rotary tool (most often referred to by the brand name of Dremel, the category leader) is decidedly low tech. It's basically a small motor with various tools that can be mounted on the motor's hub. The Dremel 8220, pictured in Figure 8.22, is a cordless model that comes with a charger and toolkit. It's not as powerful as corded versions, but it's so much handier! It's about $100 and accommodates saws, drills, polishers, grinders, and a whole lot more.

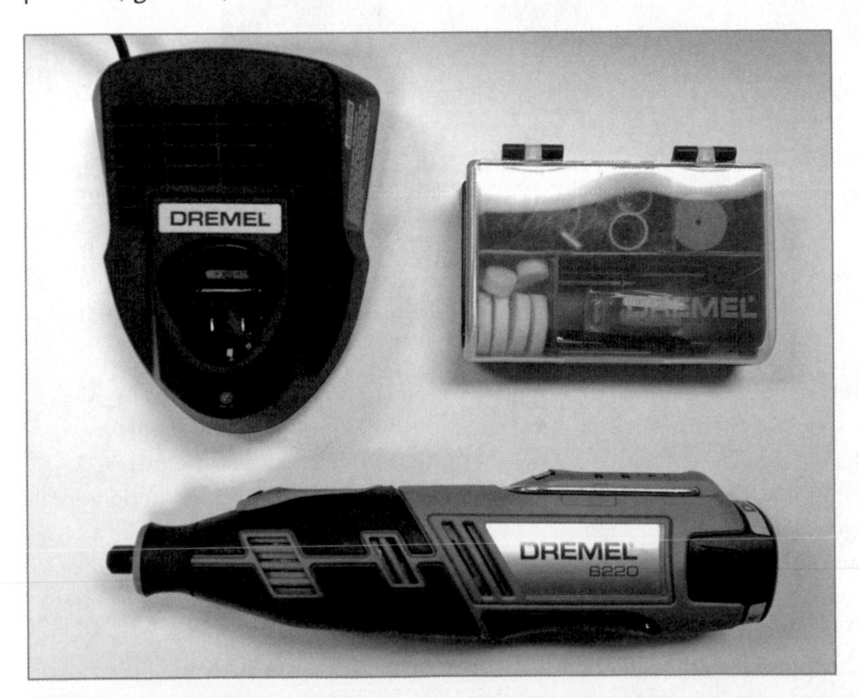

FIGURE 8.22 A rotary tool is a great way to cut, carve, and etch wood.

Air Compressor and Attachments

Many makers use air-actuated tools such as nailguns, drills, blowers, and paint sprayers. The advantage to compressed air is that the individual tools are light, because they don't need massive power supplies built in. Furthermore, you only need one compressor and can swap in any number of tools as needed. Air compressors can be dirt cheap, especially off-brand, low-capacity models (see Figure 8.23).

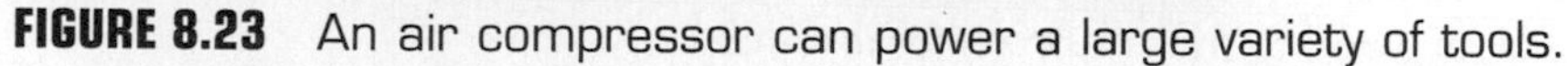

FIGURE 8.23 An air compressor can power a large variety of tools.

Drill

I probably use a drill—either handheld or a press—at least three times a week. When a drill is properly equipped with bits, you can really do a large variety of jobs with it. I use my cordless drill (an 18-volt DeWalt) at home to drive screws and bolts as well as to make holes. I use a press (pictured in Figure 8.24) to do precision work.

FIGURE 8.24 A drill press is a must-have for precisely drilling holes.

CNC Mill

A CNC mill is a computer-controlled cutter that follows a vector path (much like a laser cutter) and enables you to make cuts and grinds with great precision (see Figure 8.25). One advantage of a CNC mill over a laser is that some models can work in a third dimension, cutting into a block of wood to make a bowl, for instance.

NOTE

More About CNC Tools

In Chapter 11, "Measuring Time," you'll learn all about CNC tools, and then you make the project enclosure in that chapter from actual CNCed parts. Curious about the technology? Read more there.

FIGURE 8.25 A CNC mill carves wood with numerical precision.

Lasering and CNCing Services

Unless you actually have a laser cutter or CNC, chances are you've felt stymied when reading this book. How can you laser-cut something without a laser? Here are some options:

- Send out the files for cutting. Numerous services, such as Ponoko (ponoko.com), will accept your design files and send you back the cut pieces. Some of these services, such as Shapeways (shapeways.com), even provide 3D-printing services where you can design an object using 3D software, and the service prints it in three dimensions and mails it back to you. I talk more about 3D printers in the next section, "Working with Plastic."

- Find a hackerspace or makerspace. These are communal workshops where you can go to use their expensive tools. I talk more about this scene in "Maker Spaces," later this chapter.
- Often, educational institutions such as community colleges and even neighborhood libraries are building fabrication shops with laser cutters and CNC mills available for use. Look into it!

Table Saw

A fixture in wood shops since Grandpa's day, the table saw (see Figure 8.26) is a must for cutting large pieces of wood very quickly. It's also very likely the most dangerous tool in any woodshop, so make sure you've been taught how to use it properly. There's a product called Saw Stop that puts the brakes on the saw blade if it touches skin; look into this if you buy a saw.

FIGURE 8.26 The business end of a table saw.

Lathe

I cover lathes in greater detail in Chapter 9, "Ultrasonic Detection," but here's an overview: They're powered mills (see Figure 8.27) that rotate pieces of wood or metal, and you use lathe chisels to work the material as it turns. You can make decorative table legs with a lathe, for instance. They're also another dangerous item in the woodshop, because the rotating spindle can wrap up hair and sleeves instantly, with an injury or fatality possibly in store.

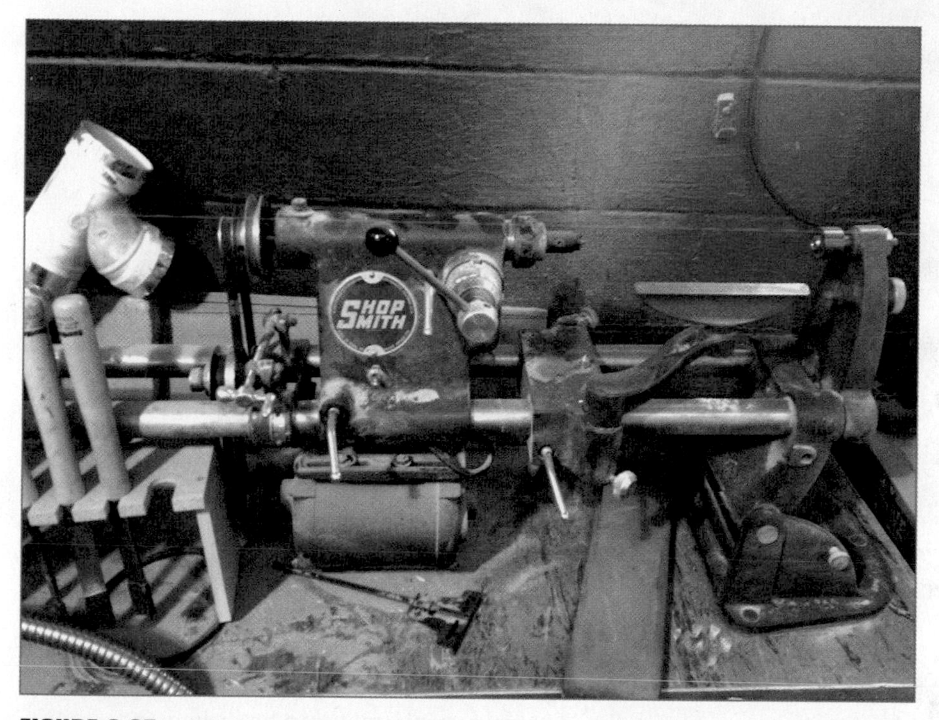

FIGURE 8.27 A lathe rotates a piece of wood, allowing you to work it with chisels.

Sander

Smoothing out rough-cut wood involves using a disk or belt sander—or like the one in Figure 8.28, both at once! As with most woodshop tools, the available options cover a wide gamut of price and function, and you'll have to choose the ideal one based on your unique needs.

FIGURE 8.28 A combination disk and belt sander allows you to smooth wood in two different ways.

Working with Plastic

Plastic can also be a very versatile medium for maker projects. It can be melted and extruded, sawed and drilled. You can print objects in plastic (see Figure 8.29) from 3D designs on your computer. You can bend it with a heat gun, as you might have read about in Chapter 4, "Setting Up Wireless Connections."

FIGURE 8.29 These robot arms were 3D-printed out of plastic.

It can also be cut into perfect shapes with a laser cutter. The laser loves acrylic! When used with the correct settings, the laser slices through perfectly, leaving polished edges.

However, you already read about laser cutters a million times in this book. What else can you do with plastic? The following sections cover some other ways you can play with the material.

3D Printers

One of the most unique and exciting developments in the realm of working with plastic are 3D printers (see Figure 8.30), which extrude molten plastic in precise paths, similar to the way CNC mills and laser cutters follow paths. However, where those tools cut away material, the 3D printer adds it. The printer creates 3D objects by extruding layer after layer of plastic until the object is formed.

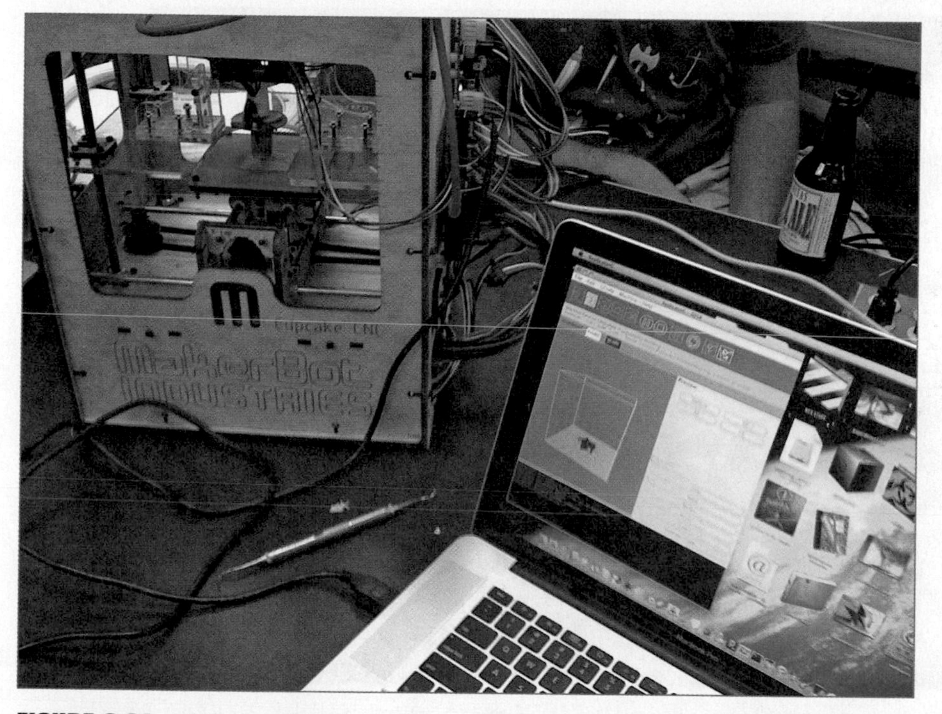

FIGURE 8.30 A Cupcake CNC 3D printer, manufactured by MakerBot Industries.

It used to be that 3D printers were the domain of successful design studios and industrial design shops. However, in the past five years, hobbyist 3D printers have been developed and have spread around the world, costing far less—but featuring inferior quality—than the professional models.

One of the most successful companies selling 3D printers is MakerBot Industries (makerbot.com), which created the Cupcake CNC printer you see in Figure 8.30. MakerBot almost singlehandedly turned 3D printing into a phenomenon that ordinary folks have heard about. Thousands of makers, teachers, and industrial designers have desktop 3D printers on their desks, and as each generation of printer gets a little bit better, look to see the technology become even more popular.

LEGO

Why go to the trouble of printing a part if you already have a bunch of similar parts sitting in your LEGO bin (see Figure 8.31)? LEGO bricks and beams have a lot of factors in their favor:

- **Ubiquity**—How many of us have owned, or still own, bucketsful of LEGO bricks? This means that if you needed to, you could probably build yourself a project box or support framework with what you have lying around.

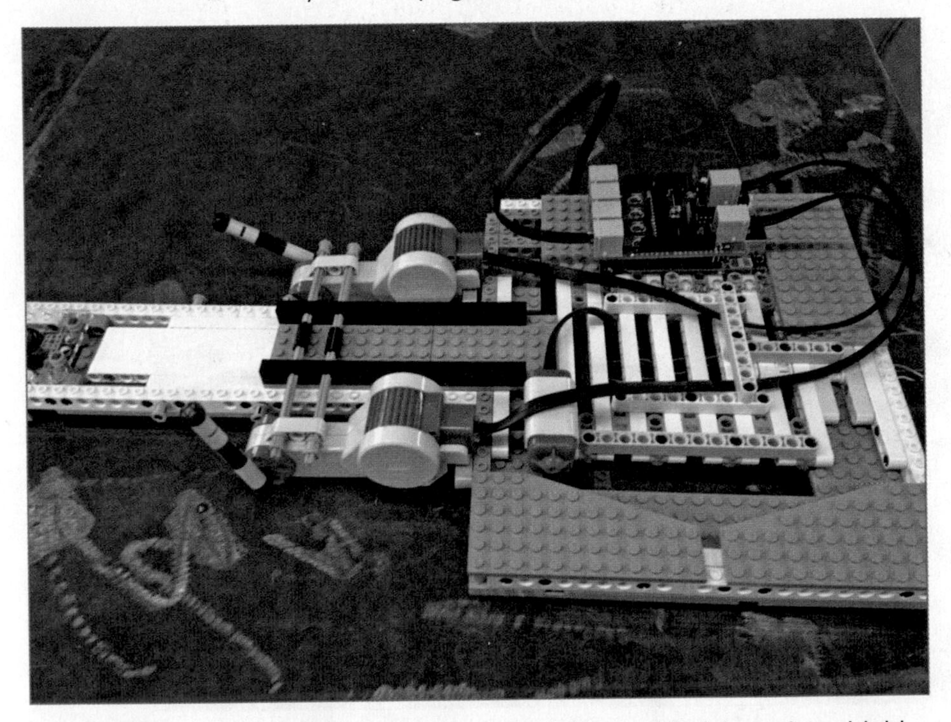

FIGURE 8.31 A LEGO "keytar" with an Arduino and Bricktronics shield controlling it.

- **Durability**—LEGO bricks are molded out of ABS plastic, pretty much the best consumer-grade plastic around.
- **Robotics**—The LEGO Group takes its robotics kits seriously, and many engineering and robotics curricula start their instruction with LEGO robotics. With motors, wheels, and sensors galore, it's hard not to be tempted.

- **Add-Ons**—Many third-party companies have developed products that can add functionality to LEGO robotics. For instance, Wayne and Layne (wayneandlayne.com) have built an Arduino shield with LEGO-compatible plugs, allowing you to control your LEGO robot with an Arduino.

Sugru

I've specified Sugru (see Figure 8.32) a few times in this book. To recap, it's a plastic modeling clay that sticks to everything and cures hard in 24 hours. You can use it to glue two things together, to reinforce or patch broken things, and to add rubber padding to tool handles. There are even makers who mold Sugru into rubber parts for their robots. An assortment of eight packets of Sugru costs $18 plus shipping.

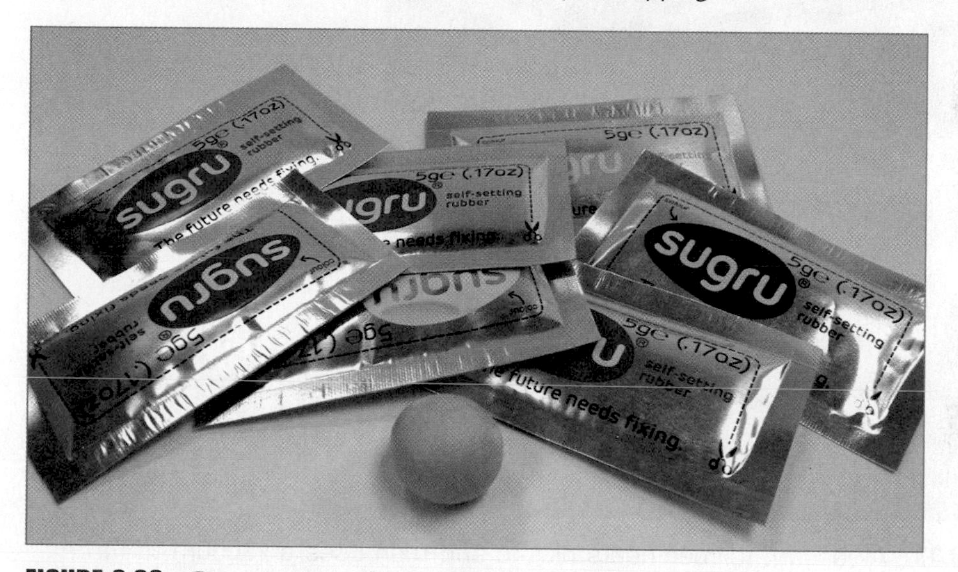

FIGURE 8.32 Sugru comes in a variety of colors.

Vacuum Former

One clever way to shape plastic is to heat it, then suck it down with a vacuum so that it hugs the shape of another object. When the plastic cools, it keeps the shape of the object. The resulting plastic shells can be painted, used as casting molds, and more. A vacuum former, seen in Figure 8.33, is a machine designed to both heat plastic as well as to form it.

FIGURE 8.33 A vacuum former heats plastic and then uses a vacuum to force the material to conform to the shape of the object being duplicated.

Extruder

An extruder (see Figure 8.34) heats plastic and forces it into molds. To make it work, you must have a mold already made; this can be a challenge in itself. You heat up plastic pellets until they're molten, and then force the plastic into the mold, where it rapidly cools. Have you heard of a Mold-A-Rama machine? It's a coin-operated, plastic-molding machine that works much the same way as an extruder.

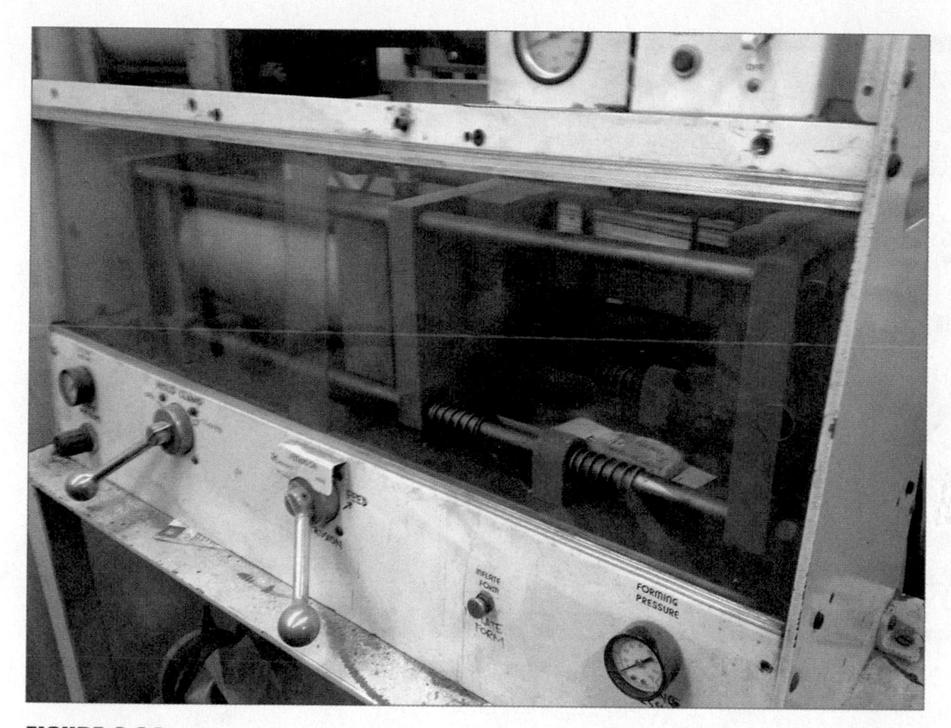

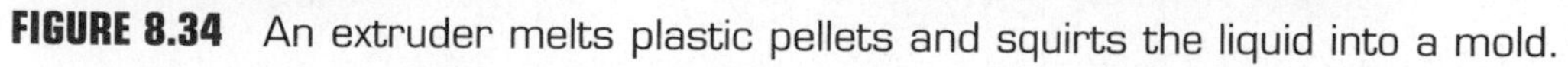
FIGURE 8.34 An extruder melts plastic pellets and squirts the liquid into a mold.

Tamiya

A Japanese hobby company, Tamiya, builds plastic robot parts such as the tank tracks in Figure 8.35. Using Tamiya (as well as other plastic robot sets) radically reduces the amount of time it takes to concept and build a robot. Want a gearbox without the hassle of designing and troubleshooting one? Go Tamiya.

FIGURE 8.35 This Tamiya tank tread kit offers a pre-made solution to designing your own.

Working with Metal

Although it's more intimidating than working with either wood or plastic, working with metal (see Figure 8.36) can be extremely rewarding as well as offer more durable results than those other materials. In this section, you'll learn about a number of metalworking tools.

FIGURE 8.36 A metal shop's welders stand ready.

Plasma Cutter

Laser cutters are great for burning through wood and plastic, but metal? Not so much—consumer-grade lasers simply aren't powerful enough to cut through metal. The solution is a computer-controlled cutter that uses plasma, or really hot gas, to burn through the metal (see Figure 8.37). If you want to precisely cut metal, this tool is for you.

> **NOTE**
>
> **Plasma Cutting Options**
>
> Note that some plasma cutters are hand-held while others use motors for control. Use the right tool for whatever project you're working on.

FIGURE 8.37 A plasma cutter uses an arc of white-hot plasma to cut through metal.

Band Saw

Just as you have band saws in the woodshop, you're likely to encounter a metal-cutting band in a metal shop (see Figure 8.38). The saw's blade is horizontal and is lifted by hand, then lowered down on to whatever is being cut. Meanwhile, a lubricant is sprayed on the cutting surface to keep the saw blade from overheating. The band saw is mostly for cutting through rods and thin pieces of metal, rather than thick ones.

FIGURE 8.38 The metal-cutting band saw is great for cutting through thin pieces of metal.

Grinder

Grinders are great for removing small amounts of surface material on a piece of metal (see Figure 8.39). Corrosion or paint, for instance, could be ground off. Grinders can also be used to shape metal, or even to cut through it.

FIGURE 8.39 Grind the surface of a piece of metal to get rid of imperfections.

Welder

The classic metal-worker's tool, welders are great for bonding two pieces of metal together. There are three major types:

- **Stick welder**—Also known as SMAW (Shielded Metal Arc Welding), this is the most basic of modern welding techniques. The welder creates an electric arc between the electrode and the surface to be welded. A consumable electrode burns and gives of vapors of inert gas, which protects the integrity of the weld.
- **MIG welder**—MIG stands for "metal inert gas" and it works by generating an arc of electricity on a joint and spraying it with inert gas (hence the name) to keep the joint free of atmospheric gas, which forms oxides and ruins the strength of the joint. The welder's gun automatically advances a spool of wire to form the weld. See Figure 8.40.

FIGURE 8.40 A MIG welder awaits the next use.

- **TIG welder**—This kind of welding uses a non-consumable tungsten electrode (TIG stands for Tungsten Inert Gas Welding) to protect the weld area with inert gas. Like the MIG, the TIG also advances metal wire to fill in the gaps of a weld.

Aluminum Building Systems

Sometimes you don't need to cut, shape, or weld metal in order to use it. Aluminum building sets allow you to build structures much the same way LEGO can, but with a great deal more strength—but they are also more expensive. The following sections cover some of the most commonplace sets.

80/20

The beams professionals use are 80/20—they are even called the Industrial Erector Set. The 80/20 (8020.net) beams come in a multitude of sizes and configurations, depending on where along the beam you want to connect other hardware such as other chassis parts, like the CNC router shown in Figure 8.41.

FIGURE 8.41 A CNC router's 80/20 beam has plastic chassis parts screwed into its T-slots.

The critical architecture of the 80/20 beam is the T-slot, which is a T-shaped groove along the length of the beam. Nuts and bolts can be attached anywhere along the slot, enabling you to build impressive structures out of multiple beams.

MicroRAX

A smaller but nevertheless very useful aluminum T-slot system, MicroRAX (see Figure 8.42) was invented in a Seattle warehouse by identical twin brothers. You can buy the beam from their store (microrax.com) or you can buy them from SparkFun. MicroRAX beams are much slimmer than 80/20, with a width of 1 cm (.4") versus 80/20's 25mm (1") and 40mm (1.5") widths. They're also much cheaper!

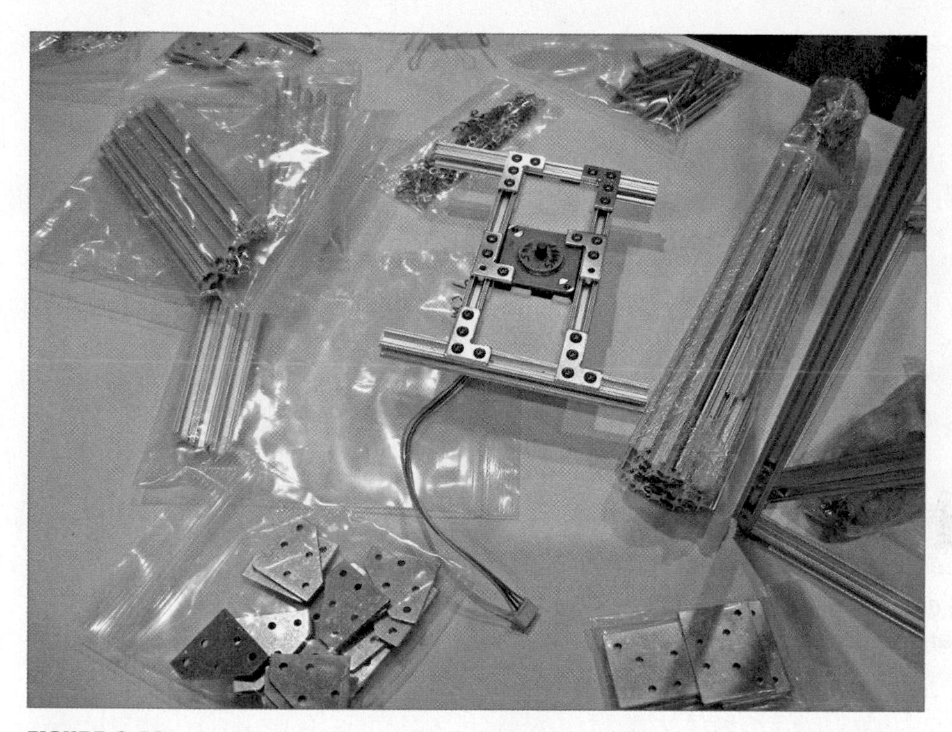

FIGURE 8.42 A MicroRAX framework supports a stepper motor.

OpenBeam

What if you made a T-slot system that followed the open source hardware ethos? That's the idea behind Open Beam (see Figure 8.43), which is designed for ease of use. The slots are 100 percent compatible with hardware-store nuts and bolts, while the slot's width accommodates 3mm Baltic birch panels. As an open source company, OpenBeam shares all technical details of its product so you can contribute to an ecosystem of hacks and innovations. You can buy OpenBeam from Adafruit, among other stores.

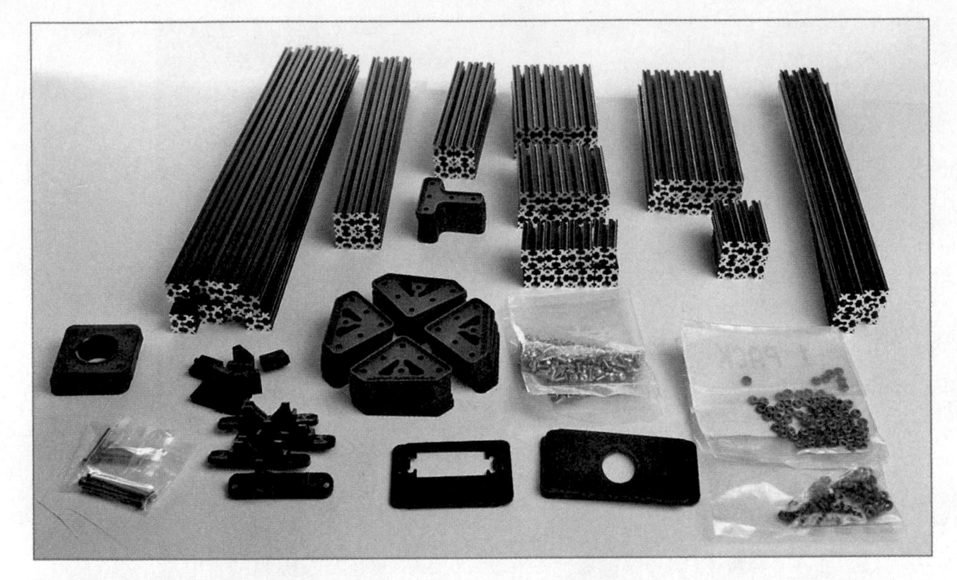

FIGURE 8.43 A variety of OpenBeam girders and attachments. Credit: OpenBeam.

Makeblock

The Makeblock company has taken a different approach than the T-slot systems, creating a complicated array of beams, gears, connectors, and wheels. Makeblock (see Figure 8.44) was conceived as the ultimate robot creation kit, and features many clever improvements over the T-slot guys, such as making the slots threaded so screws can be inserted without a nut. You can buy Makeblock at Seeedstudio.com—note the third "e"!

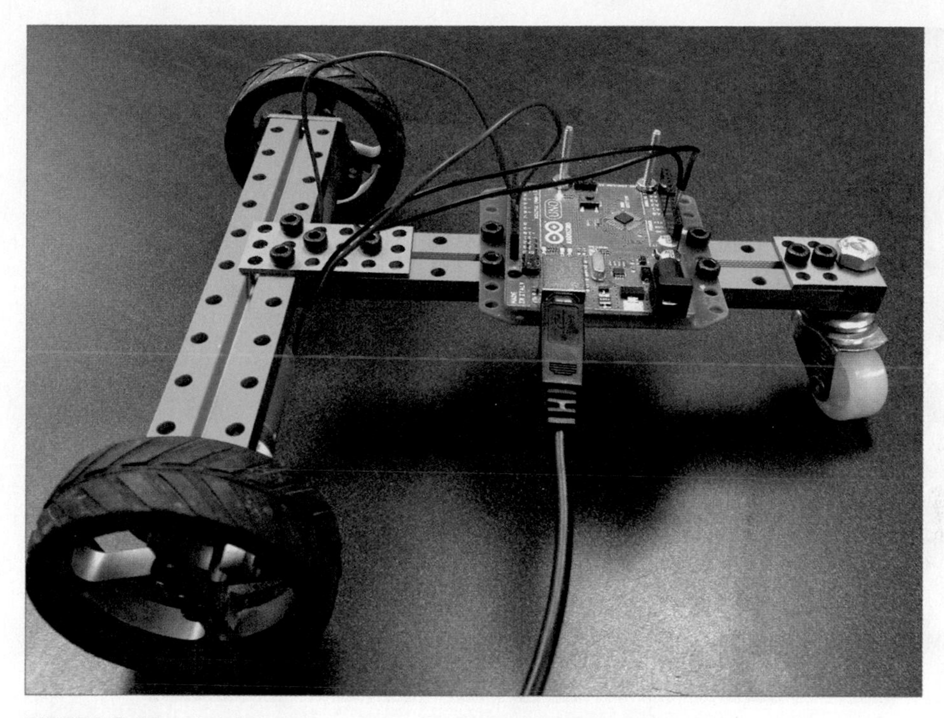

FIGURE 8.44 A Makeblock robot chassis takes form.

VEX

An educational aluminum building set, VEX (shown in Figure 8.45) is a building set like Erector with screw-hole studded metal beams held together with screws. It has its own custom microcontroller system including a wireless remote control system. You can buy VEX at vexrobotics.com.

FIGURE 8.45 A VEX robot with a battery pack and wireless receiver mounted on top.

Maker Spaces

By now you're sure to be thinking to yourself how difficult it is to have all those tools. Yes, many of them are individually cheap, but when you need a whole bunch of them, it can start to get expensive. Then there are those "big ticket" items such as laser cutters, which can cost upwards of $10,000 even for a basic model.

One solution might be a maker space (also often called a "Hackerspace"), a relatively recent phenomenon where local groups of makers rent out warehouses and pool their tools.

The Hack Factory (see Figure 8.46) in Minneapolis, Minnesota, has a full metal shop, a full wood shop, a craft area, and an electronics lab that also serves as the space's classroom. There are about 120 members, and recently (as I write this) the board approved the purchase of a laser cutter.

FIGURE 8.46 The Hack Factory in Minneapolis, complete with member-made siege machinery.

Hackerspaces are well known for their role as educational organizations. Most spaces hold regular classes (see Figure 8.47) on lockpicking, sewing, welding, and, of course, Arduinos.

FIGURE 8.47 A hackerspace's Arduino class generates recruits and money for the organization. Credit: Paul Sobczak.

The classes offer an intriguing entry into the maker arts for those not ready, or who simply aren't interested in becoming full hackerspace members. At the same time, many attendees end up joining anyway, often signing up giddily the same day as their class.

One side benefit of offering classes, beyond recruitment, is that they can make much-needed money for the organization. Frequently the money earned (classes often cost anywhere from $25 to $60) is earmarked for class-related purchasing needs; for example, using proceeds from a metalworking class to buy welding rods for the shop.

Classes aren't the only way to learn maker skills. One of the best ways is to collaborate with other makers to build a big project (such as the catapult shown in Figure 8.48) no single person could handle.

FIGURE 8.48 Hackerspace members assemble a catapult. Credit: Pat Arneson.

Maybe you have an idea for a project and don't know how to build it. You could convince another person with more skills and a little time to help you with your creation. Usually everyone brings something to the table; however, beginners are usually welcome as long as they're super interested and soak up information.

Often, special team events such as hardware hacking competitions will cause a small group of makers to band together to build a project in just a few hours or days. Usually the contests stipulate certain rules, such as only using electronics from the hackerspace's junk pile.

Still other groups band together to build products to sell, often designing electronic kits for other makers. Some of these creations end up a success, and their creators get to quit their day job and go "maker pro."

How much does this cost? A month's membership at the Hack Factory is $55, and gets you a key fob so you can access the building any time of the day or year. Other spaces are more, with some memberships upwards of $125 a month. Nevertheless, if you're bemoaning a lack of tools, you can do a lot worse than joining your local maker space.

NOTE

Learn More About Maker Spaces

Looking for a resource about maker spaces? I've written a book called Hack This: 24 Incredible Hackerspace Projects from the DIY Movement (Que 2011, ISBN 978-0-7897-4897-3) that describes two dozen hackerspaces and a project each of them is working on. It was the very first book on hackerspaces and one of the few out there that describes the culture of these groups and shares how to create your own. Check it out!

If you want to learn more, visit http://hackerspaces.org/wiki/—this is the central clearinghouse of information on the hackerspace movement.

Software

Not all tools are physical! Sometimes software can be a great help in designing electronic circuits and creating laser-cutting files. Of course, the following resources are but a fraction of everything that's out there, but the ones mentioned in this section are some of the best.

GIMP

The GIMP (see Figure 8.49) stands for GNU Image Manipulation Program, and it's a free and open-source version of the classic graphic design tool, Adobe Photoshop. It offers versions for PC, Mac, and Linux, and the menus and options are designed to resemble those of Photoshop. You can learn more about this program at gimp.org.

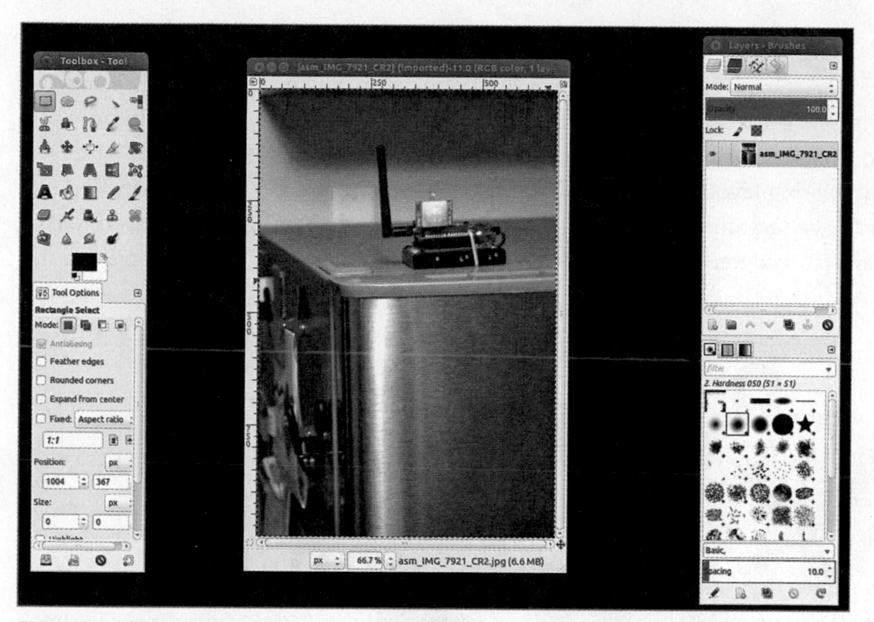

FIGURE 8.49 The GIMP is a free and open-source image manipulation program. Credit: Adam Wolf.

Inkscape

If the GIMP is the free and open-source Photoshop, then Inkscape is the equivalent to Adobe Illustrator (see Figure 8.50). It allows you to design and manipulate vector graphics, which is invaluable for generating CNC toolpaths. Files are saved as SVG (scalable vector graphics) formatted files, which is a format that most vector art programs, including Illustrator, can open.

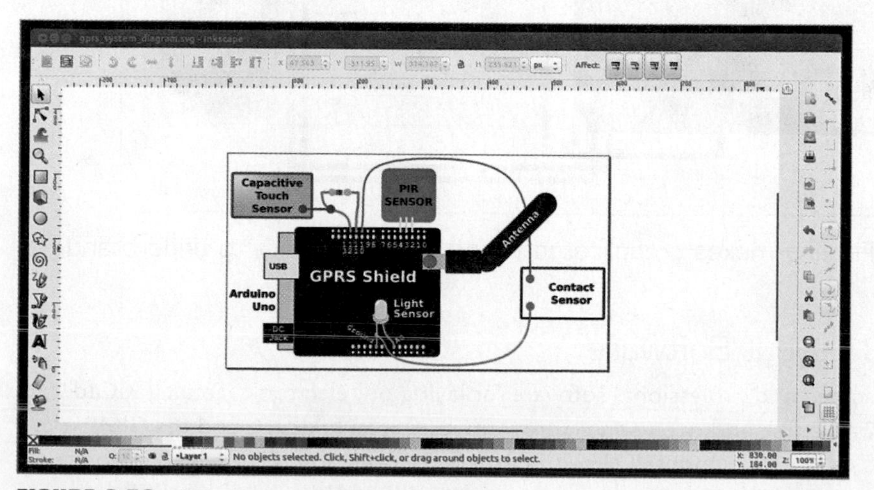

FIGURE 8.50 Inkscape allows you to create and edit vector paths. Credit: Matthew Beckler.

Fritzing

You're already familiar with Fritzing (see Figure 8.51), or at least its output. Nearly every wiring diagram in this book was generated by that program. In essence, it's the ultimate computer-based tinkerer's tool. It consists of a library of parts that can be dragged and dropped to create wiring layouts, and you can even output your design as a Gerber, the de facto format for generating printed circuit boards. That said, Fritzing is in beta, which means that it's not considered ready for official release. Nevertheless, a lot of people use it all the time.

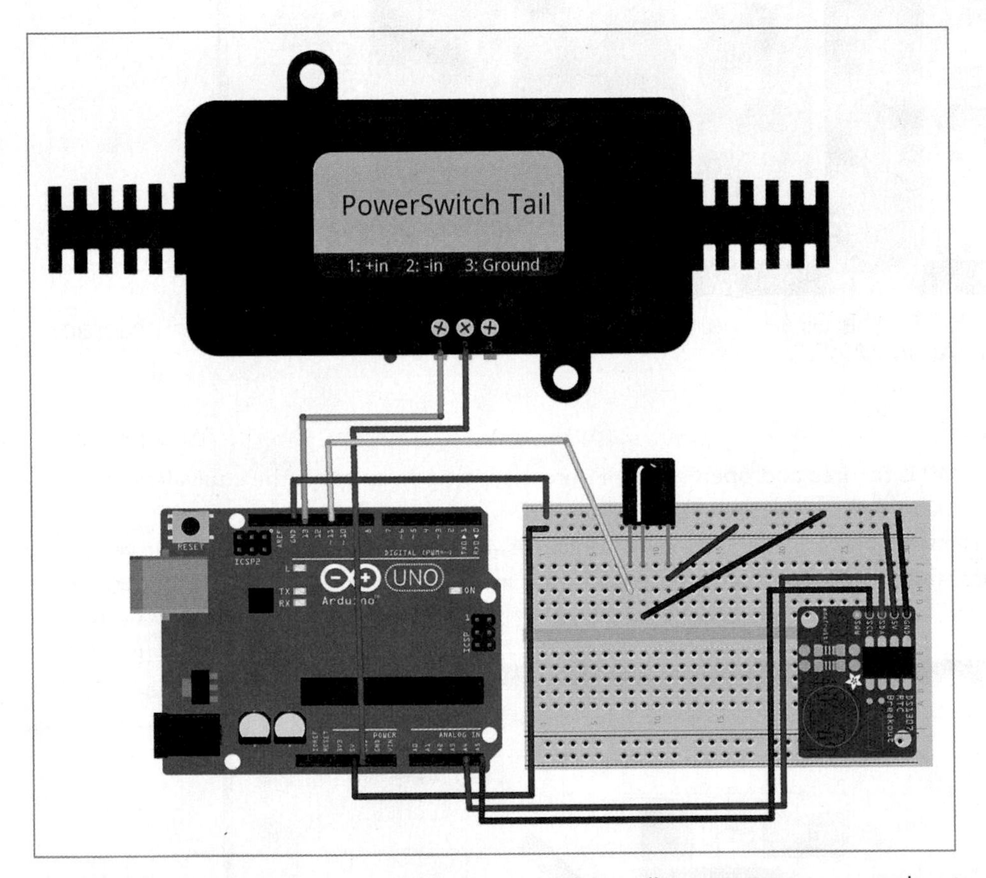

FIGURE 8.51 Fritzing makes complicated wiring diagrams easy to understand.

KiCad PCB Layout Software

A more sophisticated and professional software for laying out electronic circuits, KiCad (see Figure 8.52) is a free and open-source program much like Inkscape and the GIMP are. KiCad's focus is on designing printed circuit boards for production. This is how it works: Suppose you have created a circuit and want to make a printed circuit board (PCB) out of it. You go into KiCad, which lets you design a circuit board, route all the connections, generate Gerber files, and output a bill of materials. It more or less offers the same functionality as professional software, but doesn't cost a dime.

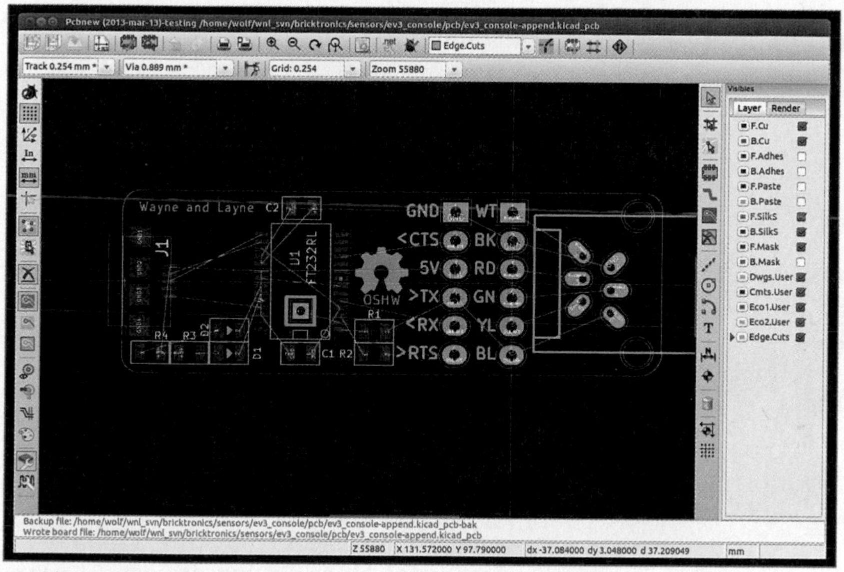

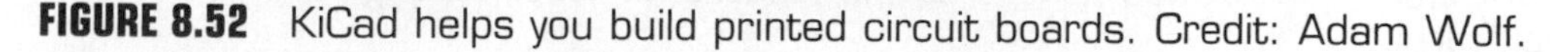

FIGURE 8.52 KiCad helps you build printed circuit boards. Credit: Adam Wolf.

MakerCase

You've seen a lot of laser-cut enclosures in this book. I created them in Adobe Illustrator because that's what I'm accustomed to using. However, what do you do if you want to create a nice laser-cut box and don't have access to a vector art program? One suggestion might be MakerCase.com, a website that generates box vectors for you so you can laser-cut all the parts (see Figure 8.53). All you do is enter your box dimensions into the site and click on a variety of options to create a box. You then download the vectors from the website. You're ready to cut out your box!

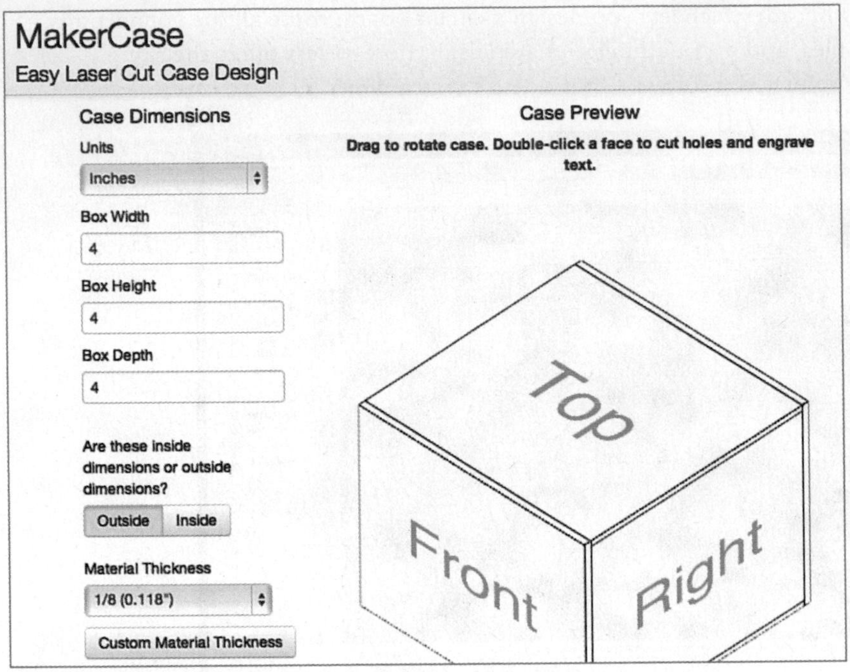

FIGURE 8.53 MakerCase generates the vectors for laser-cutting project enclosures.

Electronics Tools and Techniques

This is an electronics book, so it's only fair to include electronic tools in the tools chapter. Let's begin with that most useful of all tinkerer's assistants, the multimeter (see Figure 8.54). I then cover a couple of other non-Arduino microcontrollers as well as Arduino add-on boards, how to salvage components from junk consumer components, and a bunch of other fun stuff.

FIGURE 8.54 A multimeter is an invaluable tool for hardware hackers.

Multimeters

I've mentioned multimeters a lot in this book, but how exactly do you use one? Figure 8.55 explains the various functions of a typical low-end multimeter. Why just low end? Because the more complicated ones could have an entire chapter devoted to them and you still would barely understand anything about them. Let's focus on an easy one:

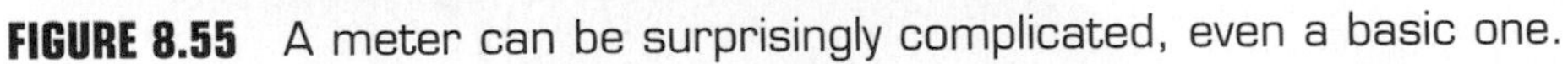
FIGURE 8.55 A meter can be surprisingly complicated, even a basic one.

A LCD screen, displaying up to four characters.

B Function selection switch. You simply turn the knob to whatever function you want.

C DC voltage. Change the switch to whatever value is closest to the value you're measuring. For instance, if you're testing a 12V battery, change it to 20V. Put one lead on the positive terminal of whatever you're measuring, and the other lead on the negative terminal.

D AC voltage. This works the same way as DC voltage. I use this meter to test outlets at home, and I set it to 200 for a 110VAC measurement.

E DC amps. Measuring amperage with a cheap meter like this one is tricky. You have to be very careful or you could ruin your meter. If you look at the selections here, you see the range goes from 200 milliamps to 200 microamps. A fuse in the meter protects it within this range, so if the amperage of the item you're testing exceeds 250 microamps, it will instantly blow.

F 10 amps. You can use this setting to measure up to 10 amps. However, unlike the current measurement I covered in callout E, this setting is unfused, so you have to be quite careful. You can only test for up to 10 seconds at a time, waiting 15 minutes between tests. If this seems ridiculous, remember that the DT-830B is dirt cheap ($10) and can't be expected to be very robust.

G Terminal jacks. A meter needs test leads to do most of its functions. There are three jacks; plug your black lead into COM and your red lead into either the top jack if you're measuring 10A, or the middle jack if you're doing anything else.

H Transistor checker. This blue plug, called an hFE socket, accommodates transistors. To test one, turn the knob to "hFE" and insert the leads of the transistor into the blue terminal based on what kind of transistor it is.

I Resistance checker. Want to know the value of that mystery resistor? Use this setting. Again, choose the value closest to the value you're testing.

J Power. Turn the knob to this setting to shut it down.

K Connectivity tester. Touch your test leads to two parts of a circuit; if they're connected, a built-in buzzer sounds.

Harvesting Electronics

You know that old Speak & Spell in the basement? Chances are it has components that you can yanked out and repurpose. The same goes for old fax machines, scanners, CD players, and other pieces of electronic junk you might have lying around.

I recently broke down an iRobot Scooba (see Figure 8.56), an autonomous mopping robot that wanders around your kitchen floor, mopping and scrubbing while you're relaxing. In addition to the expected motors and pump, the Scooba had some fascinating components such as optical proximity sensors, which detect walls and IR beacons and steer the Scooba away.

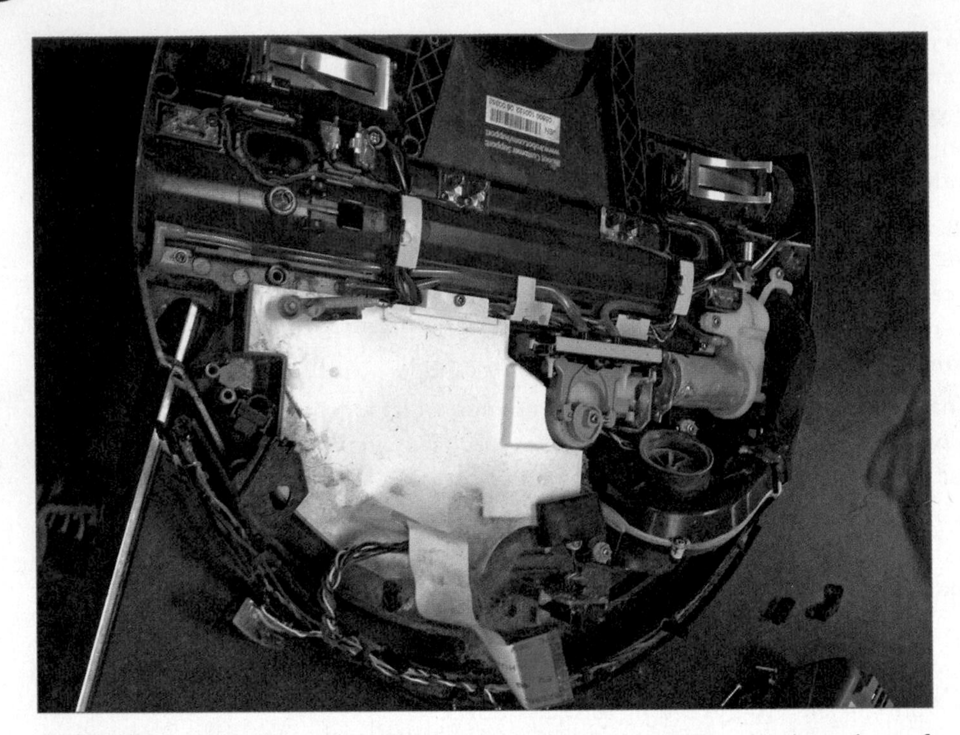

FIGURE 8.56 A broken Scooba floor-mopping robot gets broken down for parts.

Breaking down electronic junk for parts can be a lot of fun. Not only can you score cool components, but you can theoretically hack the gadget to do something different. For instance, you could swap in one sensor for another, or use a potentiometer instead of a resistor. This kind of hacking is called *circuit bending*, and the term is most commonly used when talking about cool audio hacks, such as making your talking teddy bear use a deep and foreboding voice.

One obstacle to breaking open an old piece of electronics is that sometimes—actually *usually*—the manufacturer uses obscure "security" screws such as hex, Torx, or triangle to stymie...well, who knows? Maybe they use them so kids don't wreck their toys and the company doesn't get inundated with emails from angry parents. One solution is to get every single driver bit imaginable, such as the set shown in Figure 8.57 (similar to Amazon SKU B000PLZJFK). It has a wide array of specialty bits, many of which you'll probably never use! Nevertheless, it's a sweet feeling when you suddenly realize you have all the bits you need to open up that broken toy.

FIGURE 8.57 This security bit set contains more than 100 different sizes and configurations of driver bit.

Remote control cars have motors and wheels, of course, and sometimes rechargeable battery packs and assorted switches. Best of all, if you can salvage the RC receiver and the controller still works, you could potentially add that functionality to another robot. Old tape players and boomboxes have speakers, motors, and often have cool switches that could be nabbed. Other products have piezos, battery terminals, and reusable enclosures. I once broke down a flatbed scanner and got a couple of nice stepper motors as well as some great gears and drive belts. It was a good haul!

Unfortunately, most modern electronics involve surface mount components, really tiny electronic parts that are basically printed onto circuit boards by machines because human fingers are too big and clumsy. What this means for you is that salvaging components is much more difficult because they're really small. At the same time, if you do want to, there are ways to soften the solder on those boards so the components can be scooped up.

Electronics Marking

If you find a mystery electronic part somewhere, how do you tell what it does? Sure, an LED looks different than a tilt sensor, but sometimes two radically different components look nearly the same, especially when you talk about integrated circuits, which all look like black lozenges. Even if you can recognize the component type, you still have to discern which specific part it is, because two transistors could behave differently, for instance.

The following sections cover some ways to identify which part is which.

Part Numbers

The easiest way to identify a component is to find the manufacturer's part number where it is printed on the housing. Sometimes it has several numbers, like the L293D motor driver chip pictured in Figure 8.58, and you have to discern the actual part number. In this case, the L293D is obviously the part number and the rest is some internal reference for the benefit of the manufacturer. Ultimately, all that matters is that you can figure out what you have.

FIGURE 8.58 What part is this? Try Googling the numbers printed on the housing.

Grab a component out of your parts bin and run an Internet search on all the numbers you see printed on it. Chances are, one of the numbers will give you the results you're looking for, such as a link to an electronic component seller's website or the manufacturer's data sheet.

Datasheets

Electronic components are built for engineers, and engineers like to have access to every possible bit of information so they can make a decision about which part to specify for a project. When an engineer is looking for information on a part, he or she downloads a PDF datasheet, like the one shown in Figure 8.59.

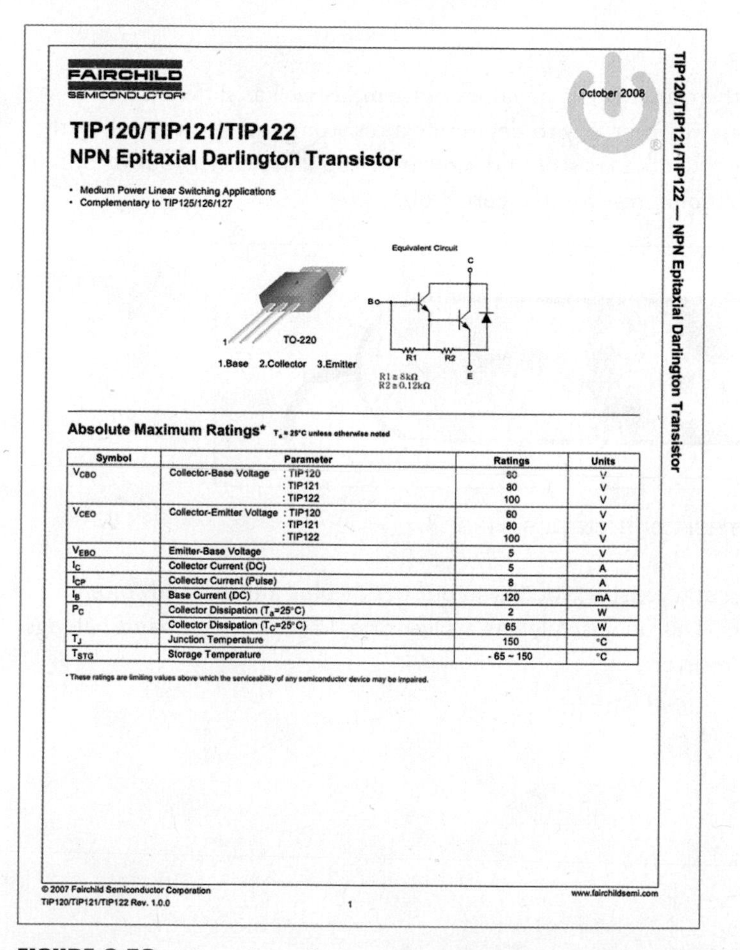

FAIRCHILD SEMICONDUCTOR®

October 2008

TIP120/TIP121/TIP122 — NPN Epitaxial Darlington Transistor

TIP120/TIP121/TIP122
NPN Epitaxial Darlington Transistor

- Medium Power Linear Switching Applications
- Complementary to TIP125/126/127

Absolute Maximum Ratings* T_a = 25°C unless otherwise noted

Symbol	Parameter	Ratings	Units
V_{CBO}	Collector-Base Voltage : TIP120	60	V
	: TIP121	80	V
	: TIP122	100	V
V_{CEO}	Collector-Emitter Voltage : TIP120	60	V
	: TIP121	80	V
	: TIP122	100	V
V_{EBO}	Emitter-Base Voltage	5	V
I_C	Collector Current (DC)	5	A
I_{CP}	Collector Current (Pulse)	8	A
I_B	Base Current (DC)	120	mA
P_C	Collector Dissipation (T_a=25°C)	2	W
	Collector Dissipation (T_C=25°C)	65	W
T_J	Junction Temperature	150	°C
T_{STG}	Storage Temperature	- 65 ~ 150	°C

* These ratings are limiting values above which the serviceability of any semiconductor device may be impaired.

© 2007 Fairchild Semiconductor Corporation
TIP120/TIP121/TIP122 Rev. 1.0.0
1
www.fairchildsemi.com

FIGURE 8.59 The datasheet of a TIP120 transistor gives engineers what they need to use the component.

You'll start collecting these sheets, more out of necessity than any requirement. You'll need them to understand which terminal does what, or to get a sense of the component's engineering tolerances. More esoterically, there's a whole bunch of information only an electrical engineer would even understand, let alone find useful.

Datasheets aren't just for individual components. Sometimes you'll see them for assemblies such as pre-soldered breakout boards packing a bunch of different parts. For instance, Evil Mad Scientist Laboratories' Three Fives Kit (P/N 652) comes with a lushly detailed spec sheet so you can delve into every aspect of the kit.

Resistor Color Bands

Resistors are tricky because there are dozens of values of them, as well as different tolerances and configurations. The best way to determine a resistor's rating is to look at the colored stripes on the housing. Grab a resistor and look at it. You'll see either four or five colored bands, like on the 470-ohm resistor in Figure 8.60.

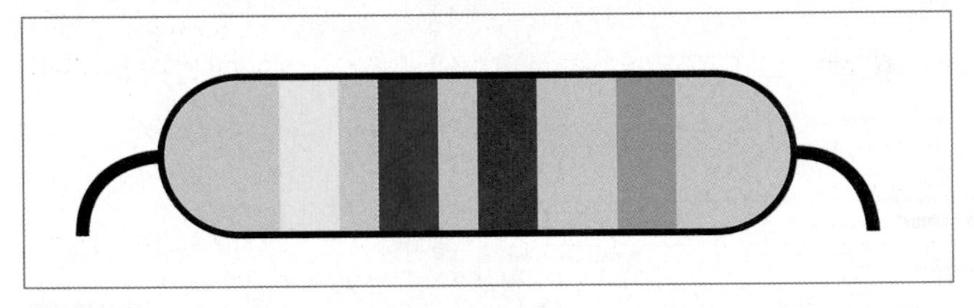

FIGURE 8.60 What do a resistor's stripes mean?

This is how it works. Looking at a resistor, you'll see four bands plus a fifth band, often slightly offset from the others. This one usually has a silver or gold band. That band belongs on the right as you read the resistor.

Each color has a number associated with it:

Black = 0

Brown = 1

Red = 2

Orange = 3

Yellow = 4

Green = 5

Blue = 6

Violet = 7

Gray = 8

White = 9

The first two bands on a four-band resistor are the base value. So in Figure 8.60, the first band, yellow, is 4 and the violet band is 7. The third band is a multiplier. That band's numerical value is actually the number of zeroes added on to the 47. Because brown stands for 1, the resistor's value is 470 ohms. If the third band had been orange, it would be a 47,000-ohm resistor.

The fourth band represents the tolerance of the component. Resistors have a tolerance, or "wiggle room" with regard to how much resistance they offer, and in projects where the resistance has to be precisely calibrated, you'll want to use a component with a low tolerance. Most resistors you'll find have a gold or silver fourth band, which represent 5 percent and 10 percent tolerance, respectively. In the case of the 470-ohm resistor you've been reading about here, the gold band means the actual value might actually fall within the range of 447 to 494 ohms.

How do you keep all these colors memorized? Neophyte engineers and makers use mnemonics, or memory aids, to keep the color bands in order. Several mnemonics are out there (you can find them on Wikipedia), but most of them are offensive or (worse) unmemorable. Here's one that you can memorize, and the first letter of each word represents each color of the resistor rating system, in order: **b**ad **b**eer **r**ots **o**ut **y**our **g**uts **b**ut **v**eggies **g**o **w**ell.

Schematic Symbols

Electronic schematics, the way engineers draw out circuitry, seems complicated (see Figure 8.61), but it's actually based on a finite number of elements pieced together.

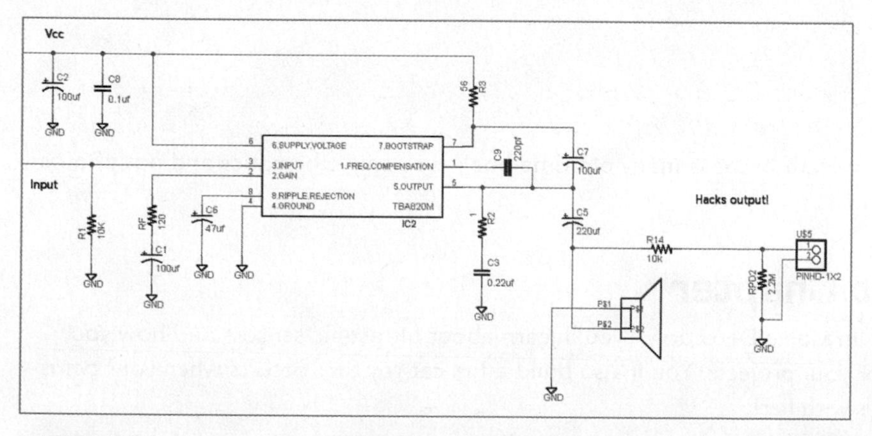

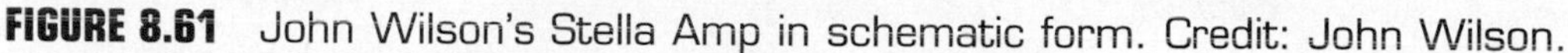

FIGURE 8.61 John Wilson's Stella Amp in schematic form. Credit: John Wilson.

Figure 8.62 shows you some of the more commonplace symbols.

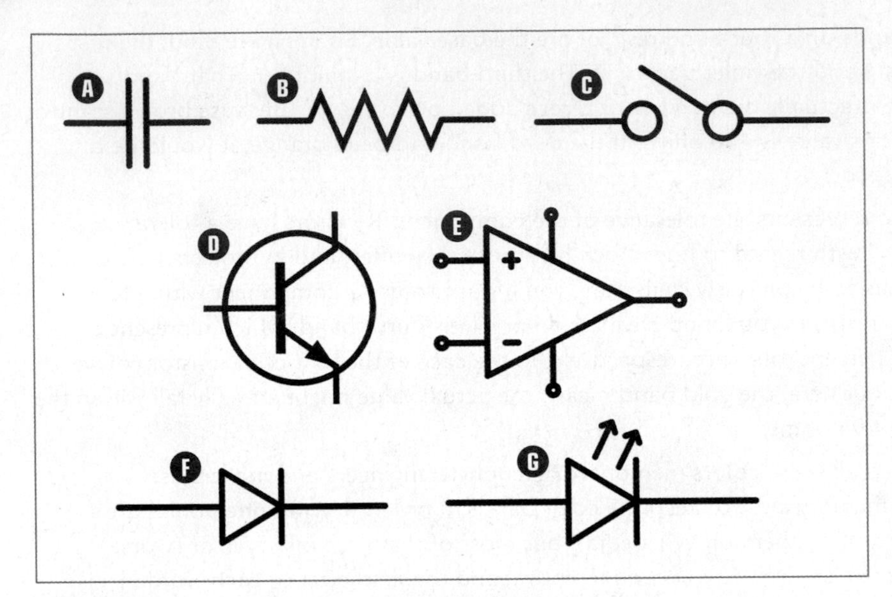

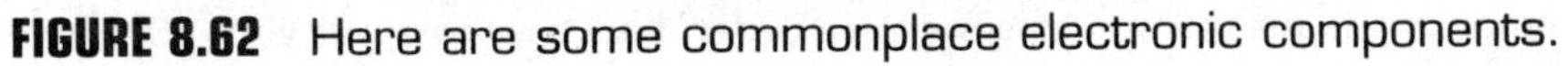
FIGURE 8.62 Here are some commonplace electronic components.

- A Capacitor
- B Resistor
- C Switch
- D Op amp
- E Transistor
- F Diode
- G LED

These are good to learn because many old-time books use only schematics and not photos to describe a circuit.

The Next Chapter

In Chapter 9, "Ultrasonic Detection," you'll learn about ultrasonic sensors and how you can use them for your projects. You'll also build a fun cat toy that detects when your pet is nearby and plays with her!

9

Ultrasonic Detection

This chapter delves into the workings of the ultrasonic sensor, an electronic module that senses the same way a bat does—with sonar. The sensor sends out pulses of inaudible sound, and then listens for them to bounce back, computing the distance traveled.

Ultrasonic sonars make excellent rangefinders, but can also be used to detect any kind of obstruction within its sensing area. Take Steve Hoefer's Tacit project (grathio.com/tacit; see Figure 9.1). It's a sonar for visually impaired people. It features a pair of Ping ultrasonic sensors paired with small servos that squeeze the wearer's wrist when an obstruction is detected.

FIGURE 9.1 The Tacit glove squeezes when it detects an obstruction.
Credit: Steve Hoefer.

After brushing up on sonar, you'll create a fun project with the technology, creating a cat's scratching post that knows when the pet is nearby and tries to play with it using a motorized dangly toy.

Lesson: Ultrasonic Detection

The sonar used for the cat toy project detailed in this chapter is the MaxBotix LV-EZ1 Ultrasonic Rangefinder, a modestly priced but robust sensor that is useful for all sorts of applications, such as range finding and detecting objects (such as panes of glass or volumes of water) that might give a light sensor some trouble.

The way it works is that the ultrasonic module can both send out ultrasonic pings, usually about 20 per second, while simultaneously listening for sound waves bouncing back—you can see this in Figure 9.2. As mentioned, this is pretty much exactly how a bat's sonar works.

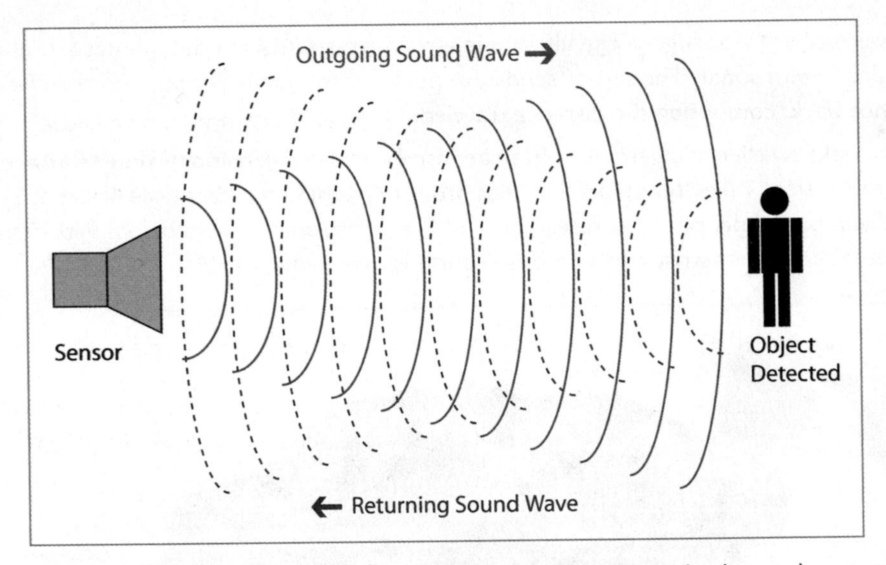

FIGURE 9.2 An ultrasonic sensor detects an object by bouncing a sound wave off of it.

Of course, this technique isn't perfect because certain textures or shapes won't reflect sound back accurately. A soft object, such as a cat or a pillow, might muffle the sound waves and not send back an accurate range, whereas a sharply angled surface might deflect the pings.

How far does the sensor detect objects? How wide a field will return an accurate result? How small an object can be seen? Uhhhhhhh... answering these questions is not so easy. Many different models of sensor are available, and they all have different sensing angles, ranges, and resolutions. The Parallax Ping ultrasonic sensor, for example, claims a range of 3 meters, sensitivity down to 2 cm, and a narrow angle of detection. The MaxBotix EZ-0 has a range of 6 meters and its sensing area is very wide, making it great for monitoring a wide area; by contrast, the MaxBotix EZ-4 has a very narrow beam, requiring that an object pass through the beam to be detected. Most beginners choose the EZ-1 because its specifications put it nicely in the middle of those extremes.

Other sensors might detect further, or have a wider angle, or be able to spot smaller objects. Be sure to check a sensor's spec sheet before you buy it, so you know what you're getting! Finally, if you're relying on your sonar for accurate rangefinding, you should also be aware that temperature variation affects sensor performance.

Ultrasonic Sensor Applications

Countless uses exist for an ultrasonic sensor. Here are some fun examples:

- Determine the water remaining in a tank by directing the sound beam down at the surface of the liquid and computing the remaining quantity based on how high the water level is in the tank.
- Create an automated store display or kiosk that activates when a customer comes near.
- Build a proximity alarm for the back of your car, buzzing when you back up too close to something. Even better, it could tell you exactly how much space you have to spare in those tricky parallel parks!
- Design a model train layout that accurately positions the train and switches tracks and opens gates accordingly.

An ultrasonic sensor is a cool gadget and great for a lot of projects! For any creation requiring accurate distance detection over short distances, ultrasonic is the way to go.

Mini Project: Make an Ultrasonic Night Light

Let's do a quick and simple Arduino project involving the ultrasonic sensor by making a light that turns on when you walk past it—in other words, a motion-activated nightlight (see Figure 9.3).

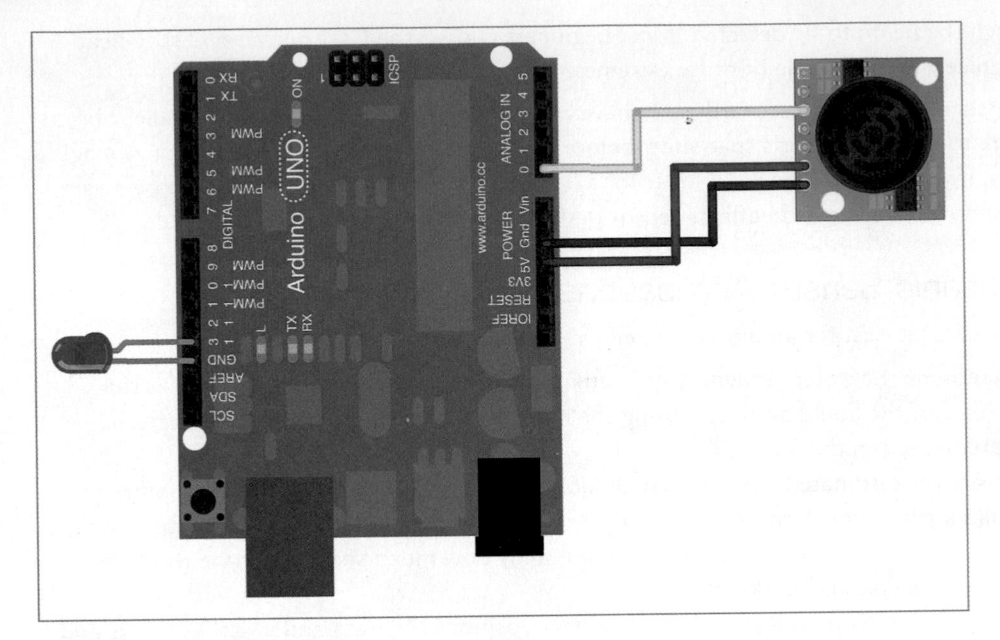

FIGURE 9.3 Seems simple? It is!

It probably won't come as a surprise to you that this is a gross overuse of an Arduino—you don't actually need a microcontroller to trigger an LED with a sensor. That said, this book is about Arduino, so you get what you get!

The way it works is that the sonar keeps its eye on the area (figuratively) and activates the LED when someone walks by. You might doubt that a single LED would make an effective nightlight, but a blue one casts a surprising amount of light! If you want, you can swap in an LED module such as a ShiftBrite (SparkFun P/N 10075, mentioned in Chapter 6, "Sensing the World") or other LED module to really illuminate the area!

Ultrasonic Night Light Code

The code is designed for using a single LED; if you decide to use a ShiftBrite or other LED module, you'll have to change the code. See the code for the main project in Chapter 6 to get an idea of how to do this.

NOTE

Code Available for Download

You don't have to enter all of this code by hand. Simply go to https://github.com/n1/Arduino-For-Beginners to download the free code.

```
int led = 13;
void setup()
{
  pinMode(led, OUTPUT);
}
void loop() {
 int distsensor, i;

 distsensor = 0;
   for (i=0; i<8; i++) {
     distsensor += analogRead(0);
     delay(50);
   }

   if (distsensor < 500) {
     digitalWrite(led, HIGH);
     delay(30000);                    // wait for 3 minutes, then recheck
  }
}
```

Project: Cat Toy

The project for this chapter involves creating a fun cat toy that interacts with your friendly local feline. It consists of a scratching post with a motorized cat toy that dangles down, giving your friend something to bat at when she's not sharpening her claws on the post (see Figure 9.4).

FIGURE 9.4 This chapter's project helps you make a fun toy for your favorite kitty.

You create the enclosure for this project—the scratching post—by creating a wood cylinder on a lathe, which is a machine that rotates a piece of material and allows you to shape it using handheld tools. It's an awesome machine to learn to use, and you'll find all sorts of uses for it.

By the way, the kind-hearted among you might be concerned that cats might be bothered by the sonar's pings. It turns out that the sonar is outside a cat's hearing range, and cats won't be bothered by it.

Anyway, let's get started!

PARTS LIST

This is a simple build with relatively few parts. This is what you'll need:

- Arduino Uno
- USB cable for Arduino with wall adapter, or you could use a wall wart
- Jumpers
- Heat shrink tubing
- MaxBotix LV-EZ1 Ultrasonic Rangefinder (Adafruit P/N 172; you can use other makes and models of sonar)
- Servo (I used a Hitec HS-322HD; see the nearby sidebar, "The Servo")
- Heavy wire (I used some welding rod, but anything on par with a wire coat hanger will work)
- Standoffs (I used 3/8" plastic standoffs, SparkFun P/N 10461)
- L-brace (Home Depot P/N 339563)
- #4 machine screw with washer and nut
- #4x3/4" wood screws
- #6x3/4" wood screws
- #6x2" wood screws
- Wood glue
- Scrap wood for enclosure (I used 1x6 pine)
- Your cat's favorite pom-pom style toy
- Drill and drill bits (7/64", 3/16", 1/2", and 1")
- Lathe
- Table saw
- Band saw
- Glue gun
- Hole saw
- Needle-nose pliers

THE SERVO

Chapter 13, "Controlling Motors," covers a lot more about cool motors—including servos—that you can use in your projects. In the meantime, let's learn about the specific motor used in this chapter's project.

The Hitec HS-322HD servo is a great all-purpose motor, the sort that you would buy if you could have only one (see Figure 9.5). That said, you can have as many as you want, so you might find the HS-322HD too slow, or too big, or not strong enough. Every motor has a spec sheet that you can download, packed with all the info you need to make a decision.

FIGURE 9.5 The Hitec HS-322HD is a great all-around servo.

Here's the scoop on the HS-322HD: It takes .19 seconds to rotate 60 degrees on 4.8 volts but only .15 seconds to go the same distance with a 6v power supply. It has 41.66 oz/in of torque (the motor's strength) on 4.8v and 51.38 oz/in on 6v. What does that mean? Essentially, it means that the HS-322HD is middle-of-the-road in many respects. If you want to get a faster servo, Hitec will sell you one twice as fast: 0.9 seconds to go 60 degrees on 4.8v. Motors with more torque are available as well.

Instructions

First, you'll wire up the electronics, and then tackle the construction of the project's enclosure, which is made out of wood and resembles a cat scratching post.

The electronics are a cinch! Just wire up your Arduino, as you see in Figure 9.6.

1. Plug in the yellow wire of the servo to digital pin 9. The black wire plugs in to a free GND pin. Hold off on the red wire; you'll be doing something special with that one.

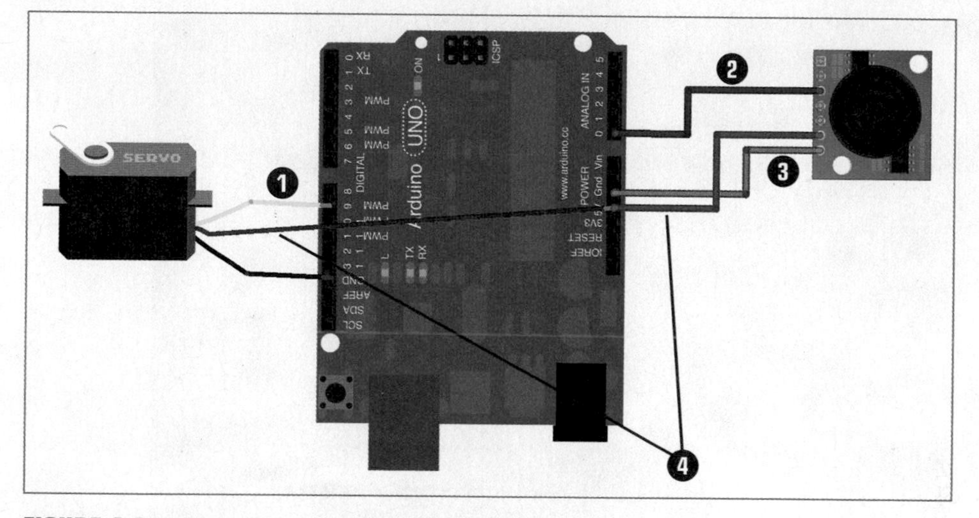

FIGURE 9.6 Wiring up the cat toy is easy!

1. Servo's yellow wire goes to digital pin 9 on the Arduino
2. Data wire from ultrasonic sensor goes to Analog 0 pin on Arduino
3. Ground wire from ultrasonic sensor goes to a GND pin on Arduino
4. Spliced wire connects the servo and the sensor to the 5V pin on the Arduino

2. Connect the data wire of the ultrasonic sensor (the purple wire in Figure 9.6) to the Analog 0 pin, and connect the ground wire (shown as gray in Figure 9.6) to a GND pin. You connect the orange wire in the next step!
3. The only tricky part of wiring up the Arduino is that both the motor and the sensor need 5V, and there is only one 5V pin on the Arduino. What do you do? You can create a spliced wire (see Figure 9.7) by soldering three wires together, and then sealing it all up with heat-shrink tubing (shown as a red wire coming from the servo and an orange wire coming from the sensor in Figure 9.6). The single end goes into the Arduino's 5V port, while the other two connect to the ultrasonic's and servo's power pins.

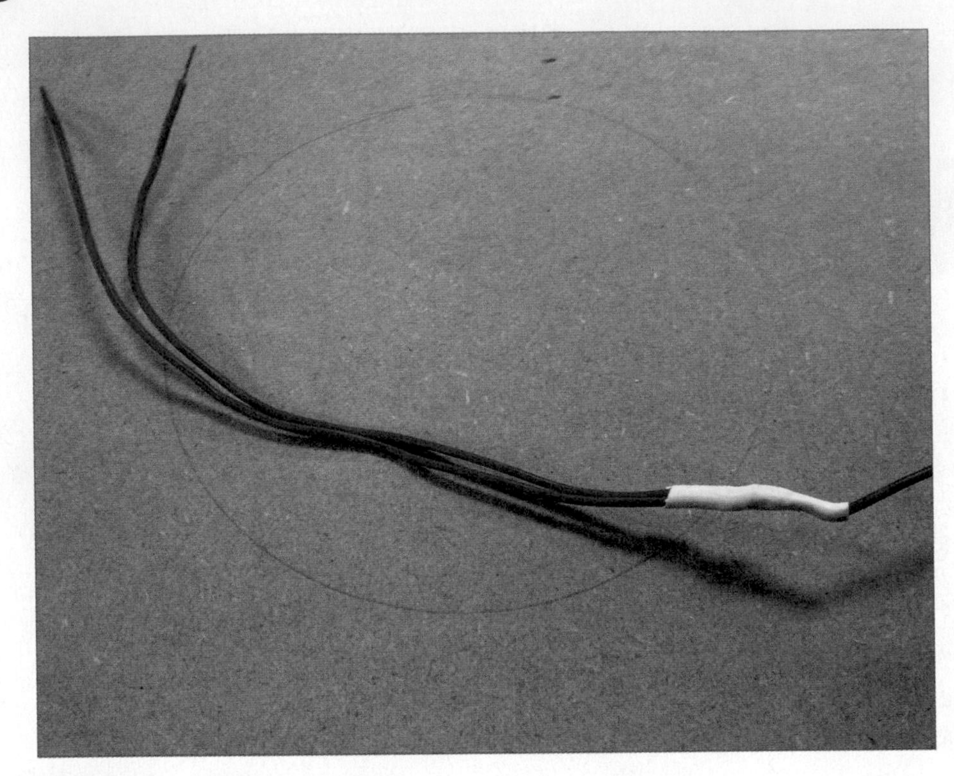

FIGURE 9.7 Splicing the wires.

The next section covers how to build the scratching post enclosure, and then you'll put it all together.

Enclosure

After you assemble the circuit, it's time to add it to the enclosure. But you have to build the enclosure first! To create the rounded shape shown in Figure 9.8, you use a lathe.

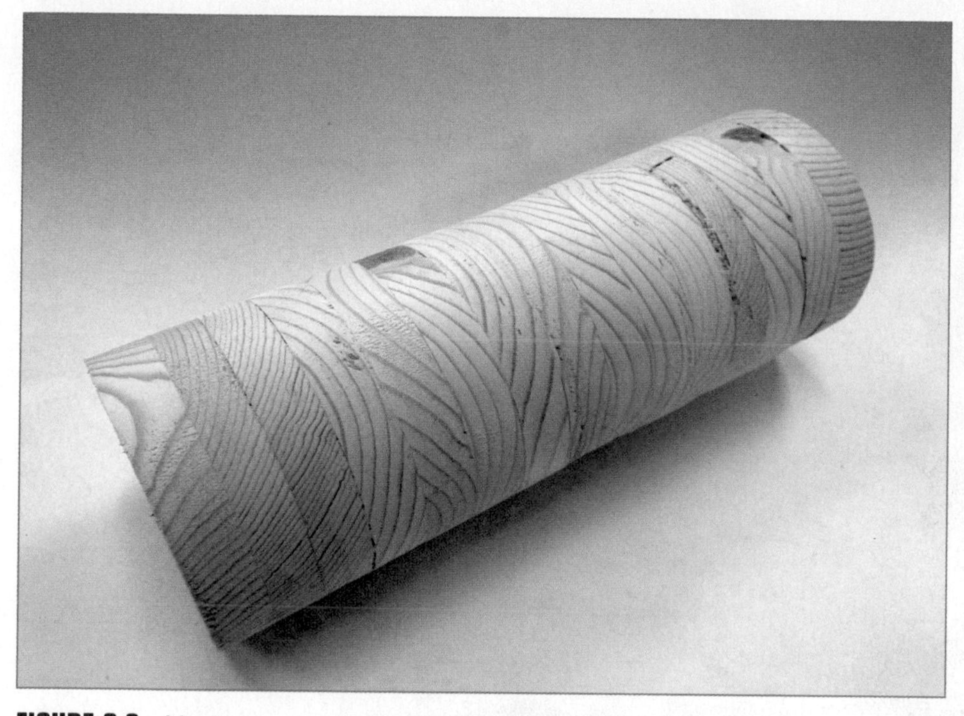

FIGURE 9.8 You can sure make a beautiful cylinder on a lathe!

For this enclosure, forget the laser cutter! You'll use old-school tech to build this enclosure (refer to Figure 9.8) out of wood. You'll cut out a bunch of rings of wood, glue them together, and then smooth out the exterior using a power tool called a *lathe*. Here are the steps:

1. Trace rings onto pieces of pine: I used a roll of tape as a template, as shown in Figure 9.9. This wood doesn't have to be great; the stuff I used was scrap wood from someone else's project. For my cat toy, I used about a dozen rings each about an inch thick, plus top and bottom plates.

FIGURE 9.9 Tracing out the rings with a roll of tape.

2. Cut out the circles using a band saw. I suggest making straight cuts into the wood (see Figure 9.10), because a band saw blade sometimes has trouble curving around a circle.

FIGURE 9.10 Cutting out the circles on a band saw.

3. Cut out the inside of the rings using a hole saw (see Figure 9.11). I probably should have cut out the insides *before* doing the outsides, because the circles wanted to spin around on the drill! I ended up using a clamp to secure the disks after blistering my fingers.

FIGURE 9.11 Cutting out the insides.

CAUTION

Don't Drill Holes in Every Disk!

Be sure to leave a couple of your disks without holes because you need solid pieces for the top and bottom.

4. Stack up the rings, reserving the two solid disks for the top and bottom. Arrange them as neatly as you can, and glue them together, as shown in Figure 9.12.

FIGURE 9.12 Glue the rings together to form the cylinder.

5. Clamp the stack of rings (see Figure 9.13) and let the assembly dry overnight. You should probably try to make your stack a little neater than I made mine, but it doesn't have to be perfect—after all, the lathe's job is to smooth it down.

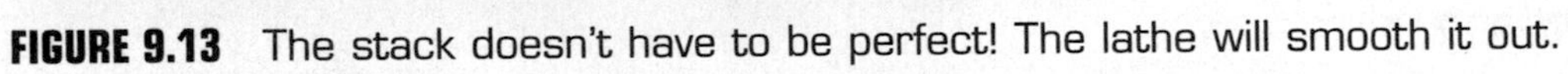

FIGURE 9.13 The stack doesn't have to be perfect! The lathe will smooth it out.

6. Put the disk stack on your lathe (see Figure 9.14). This is where having solid disks at the top and bottom come in handy. Use the lathe tools (see the later section, "Lathe 101") to smooth out the sides.

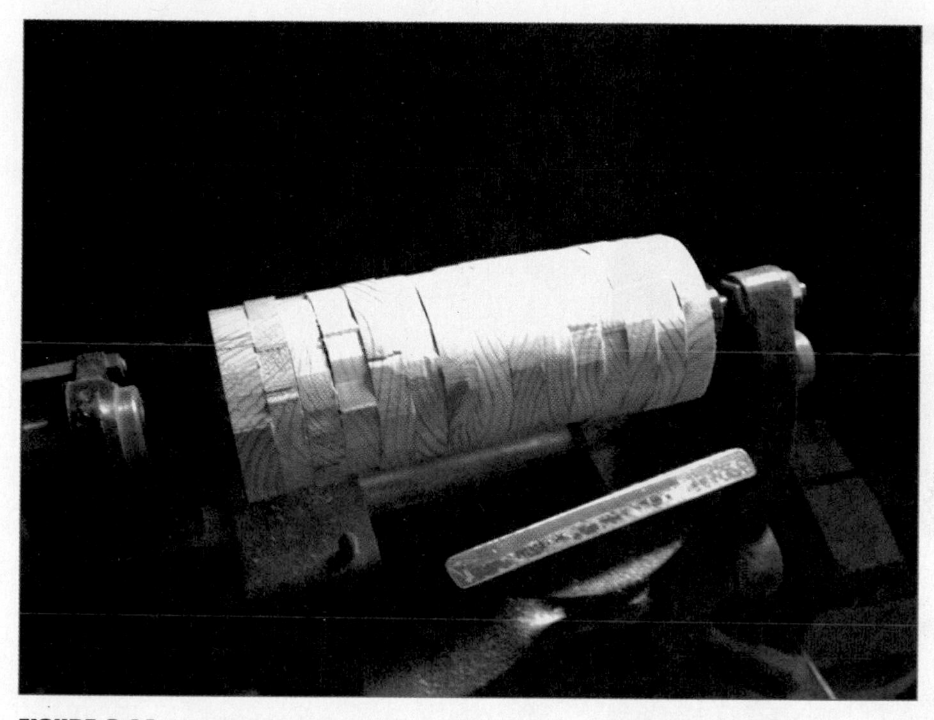

FIGURE 9.14 Putting the glued stack on the lathe.

Fresh off the lathe, the stack of disks looks great! You can see a slight taper in the middle of the cylinder, shown in Figure 9.15. I just pressed down on the tool a little too much in the middle. I could have evened it out if it really bothered me, but it didn't!

FIGURE 9.15 All finished with lathing!

7. Cut off the top and bottom of the ring cylinder to put the electronics inside, as shown in Figure 9.16, as well as to cut holes for the ultrasonic and the power supply.

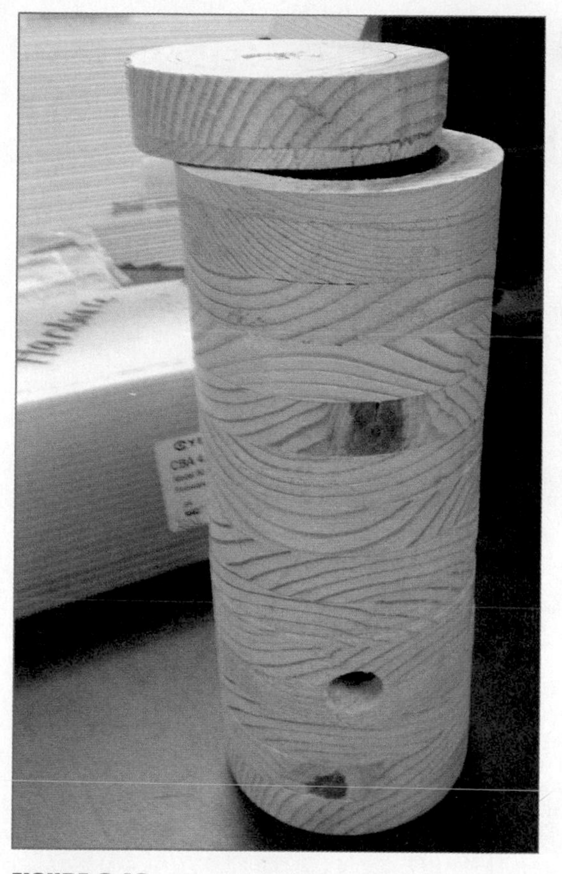

FIGURE 9.16 Cutting and drilling the cylinder.

8. Add the servo. This involves drilling a hole for the heavy wire that will dangle the cat toy. Use a 3/16" bit and drill completely through the top disk. Then drill down about a quarter inch with a wider bit—I used a 1/2" bit—to accommodate the servo's hub, which protrudes somewhat (see Figure 9.17). You'll also want to drill the holes for the hardware that connects the motor to the enclosure, and I used 1" bits for these. It'll be obvious where to drill these holes when you position the motor.

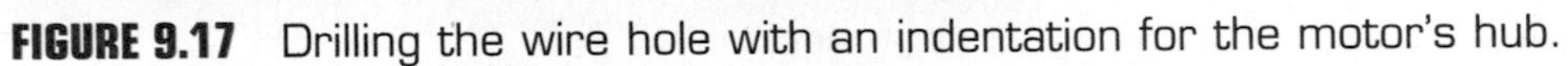

FIGURE 9.17 Drilling the wire hole with an indentation for the motor's hub.

9. Connect the heavy wire to the hub. In my case, I used hot glue but you might decide you want a more robust attachment. You then thread the wire through the 3/16" hole you drilled in the top and when it's flush, attach it with the #4 × 3/4" wood screws and 3/8" plastic standoffs, as shown in Figure 9.18.

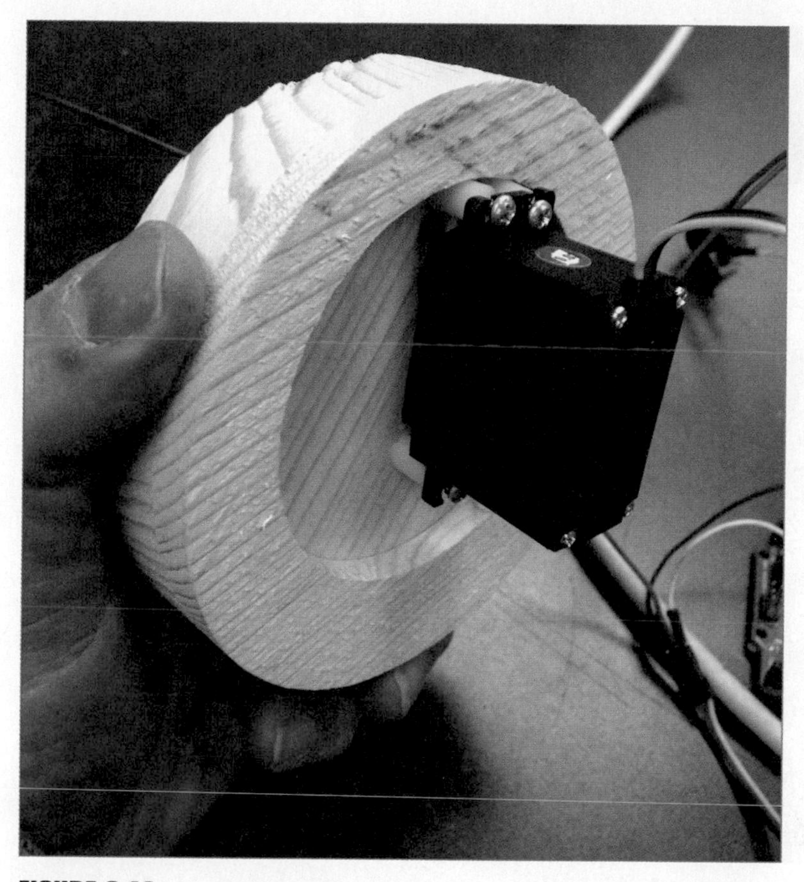

FIGURE 9.18 Attaching the motor with screws and stand-offs.

10. Insert the ultrasonic sensor through the base of the cylinder and stick it through the 1" hole you drilled (see Figure 9.19). I used hot glue to secure it; there are screw holes in the sonar's circuit board, but I found them to be fairly inaccessible.

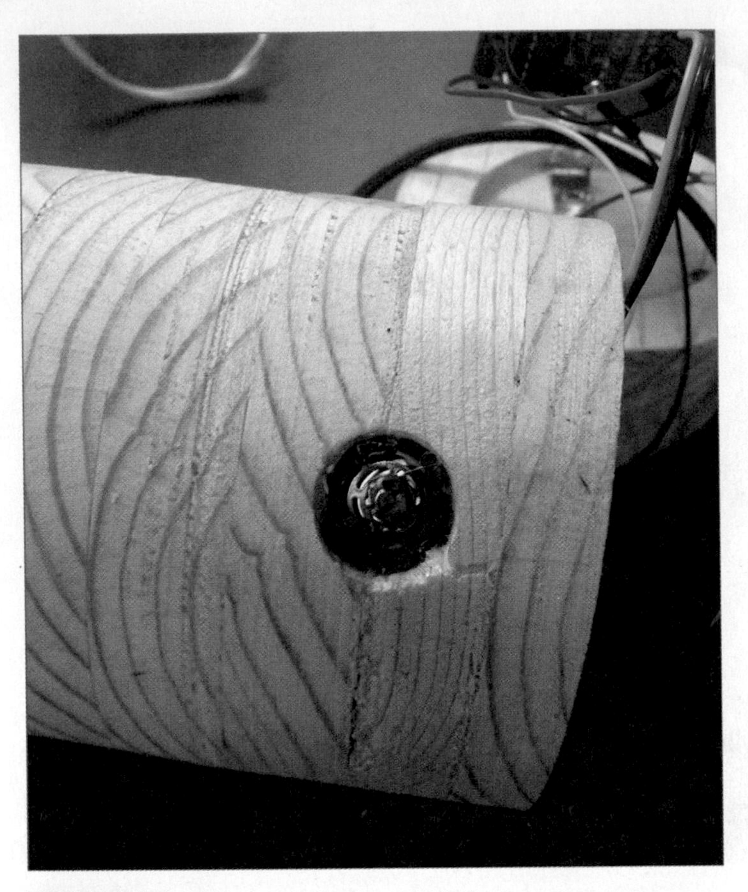

FIGURE 9.19 The business end of the ultrasonic sensor peeks out of the enclosure.

11. Attach the Arduino as shown in Figure 9.20. I used a hardware store L-brace to connect the Arduino using a #4 machine screw. How do you connect a square to the inside of a cylinder? Don't— simply connect it to one end!

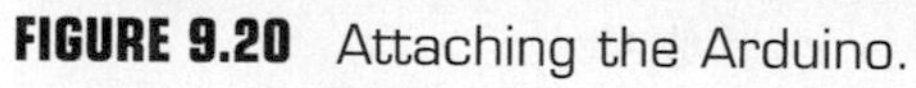

FIGURE 9.20 Attaching the Arduino.

12. After the motor is in place and everything is wired up, screw down the top (see Figure 9.21) with some 3/4" #6 screws. Do the same for the bottom.

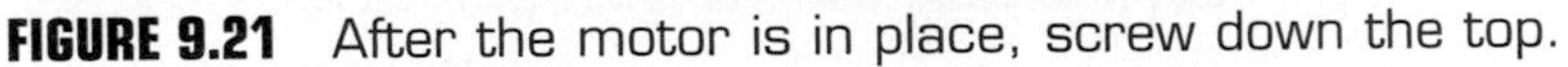

FIGURE 9.21 After the motor is in place, screw down the top.

13. Make the base. I cut a 9" × 9" square of 1.25" MDF on the table saw, as shown in Figure 9.22. It has a nice solid heft to it! Attach the cylinder to the base with the 2" #6 wood screws. One consideration to keep in mind as you do so is to be sure you don't accidentally drill into the screws used to connect the base to the cylinder. An easy way to make sure this doesn't happen is to drill into the middle of the base rather than the edges. Don't worry, you can't run into the screws holding the L bracket to the bottom disk because the wood is too thick!

FIGURE 9.22 Connecting the cylinder to the base.

CAUTION

Look Out!

By the way, do you see that weird gouge in Figure 9.22? The table saw grabbed the wood and flung it back at my head—fortunately, it missed! Working with wood is dangerous; make sure you're using your tools properly and are following all safety precautions.

14. Wrap the post in cloth so the cat can scratch on it if she gets bored with the pom-pom. I used corduroy, which might not be the best material, but it looks great! Other options might be carpet either glued or stapled to the wood or sisal twine wrapped around the cylinder. I measured the post's circumference with a flexible tape measure, and then its height. I then cut out the corduroy with a pair of scissors (see Figure 9.23). I applied wood glue to the post, and then wrapped the cloth around until it stuck. I used an X-ACTO knife to cut out the holes for the ultrasonic and the power cord.

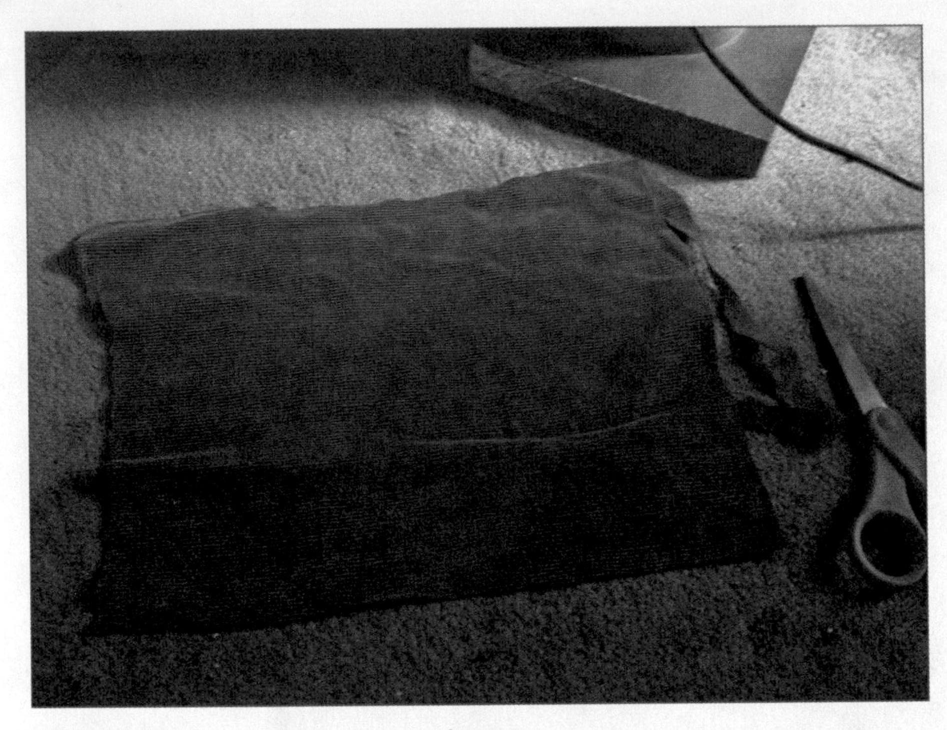

FIGURE 9.23 Preparing to wrap the post.

15. Twist a loop at the end of the heavy wire with a pair of needle-nose pliers and connect the pom-pom to the loop. You're finished!
16. Find a cat to amuse (see Figure 9.24).

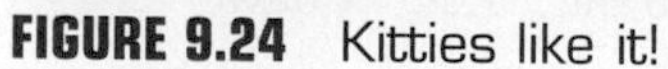
FIGURE 9.24 Kitties like it!

Lathe 101

A lathe (see Figure 9.25) is essentially a motor that rotates a piece of wood or metal on its axis so that it can be worked on with a carving tool or sanded, polished, painted, or anything else. Some lathes have an attachment allowing the inside of a cylinder to be bored out; unfortunately for this project, the lathe I used doesn't do that.

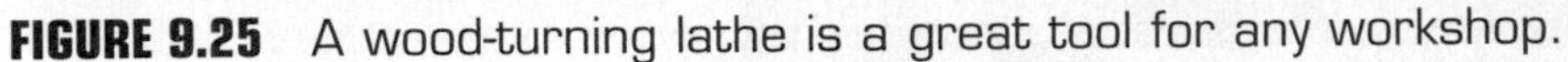
FIGURE 9.25 A wood-turning lathe is a great tool for any workshop.

So, what use is it to spin something that you want to carve? Basically, you can use it to make beautiful cylindrical objects such as table legs and candlesticks. Although many different types of lathes are available, this discussion pertains to the classic woodworker's tool.

Here's how to work an object on the lathe:

1. Prepare the item. If you can shape it reasonably smooth with hand tools, you'll save time on the lathe.
2. Connect the item to the lathe. You want it as centered as humanly possible. You can either screw connector plates to the wood, which ensures that you have it perfectly centered, or you can use a *mandrel,* sort of a tooth that pokes into the wood and secures it.
3. Spin the item on the lathe, and work it with woodworking chisels (see Figure 9.26) until it looks the way you want it.

FIGURE 9.26 You use long-handled woodworker's chisels to carve into the spinning wood.

If you want to learn more about lathes and how to use them, I suggest doing a YouTube search on "how to use a wood lathe" or something similar.

Lathe Safety

When using a lathe, follow these very important safety rules:

- Make sure an experienced operator gets you "checked out" on the lathe; in other words, someone has walked you through the machine's functions.
- Wear ear and eye protection whenever the machine is in use.
- Make sure all the lathe's adjustable parts are secured before you start the motor, and the mandrel (if used) is firmly seated.
- A lathe can bind up loose items such as hair and sleeves. This can potentially be fatal, so keep your hair up and avoid free-flowing clothing such as ties and puffy sleeves.

The Next Chapter

In Chapter 10, you'll learn about ways to make cool electronic noises with your Arduino. You'll build a sweet handheld noisemaker that generates a multitude of crazy sounds.

10

Making Noise

You can do a lot of crazy things with an Arduino, and one of them is making noise! All you really need is a speaker wired in to a couple of pins, but you can add fun extras like buttons, knobs, and sensors to modify the sound. In this chapter, you'll examine a few ways you can generate cool sounds with your Arduino, and then build a fun noisemaking toy (see Figure 10.1).

FIGURE 10.1 In this chapter, you get to build a cool noisemaker.

Noise in Electronics

People have been making electronic music ever since electronics were invented, and some of the most adventurous and creative of these folks are ordinary people hacking at home.

Take the phenomenon of circuit bending, for instance. Circuit bending (see Figure 10.2) involves modifying existing electronic noisemakers, such as a Speak & Spell toy. Circuit benders dismantle the gadget and play around with the electronics to create cool sound effects.

FIGURE 10.2 Mickey Delp tinkers with his circuit-bent caterpillar toy. Credit: Pat Arneson.

One way they do this is by replacing a key resistor with a potentiometer, also known as a variable resistor. Turning the potentiometer's knob changes the resistance, and might also change the way the toy sounds. Other tactics involve swapping in different capacitors, adding timer microchips, and even adding an Arduino or other microcontroller for more detailed control of the toy's various effects.

Circuit-bending is not the only kind of electronic music, however. Setting aside professional noisemakers as well as professional music applications, a lot of cool projects are still out there. The basic premise of many of them is to generate a tone, while using manual input like potentiometers and buttons to modify the sound.

Some of these projects are so successful that they've actually been turned into commercial products, often in kit form, meaning that you have to assemble it yourself. Let's look at a couple of noisemaking projects that have been turned into products.

Thingamagoop

Austin, Texas-based circuit benders Bleep Labs created the Thingamagoop (see Figure 10.3). It features switches, knobs, and a button, as well as a light sensor and LED antenna. The Thingamagoop features sample and hold, arpeggios, noise, and bit crush effects.

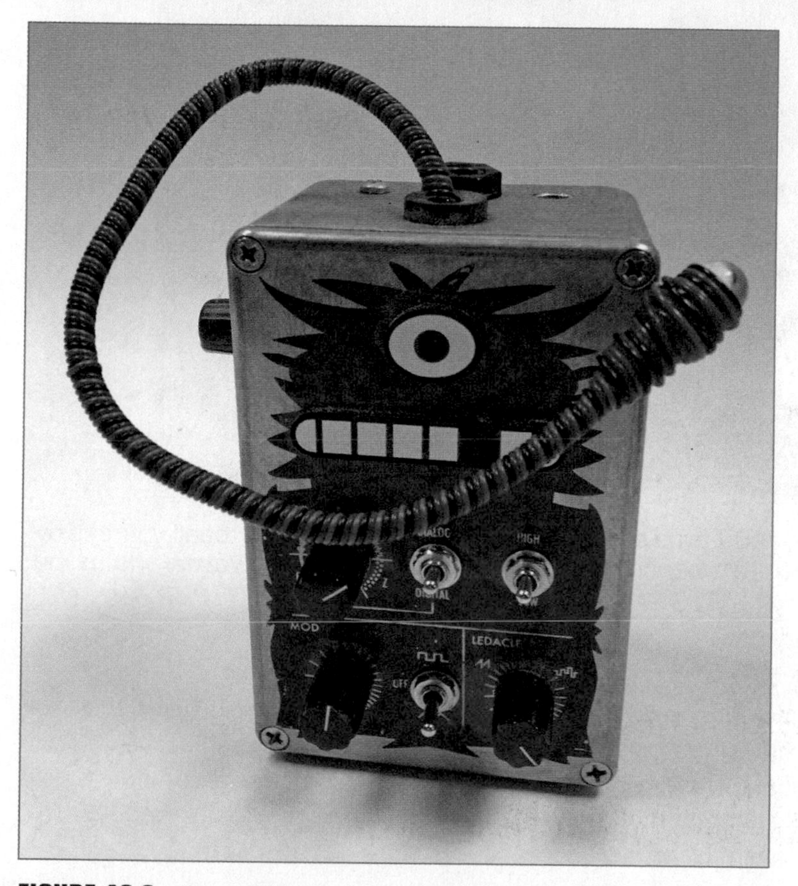

FIGURE 10.3 The Thingamagoop looks cool and makes even cooler noises (Bleeplabs.com, $120).

Tactile Metronome

Electronic kitmakers Wayne and Layne built the neat kit shown in Figure 10.4. You tap on the piezo buzzer with your finger, and the microcontroller detects the vibration and records the pattern you tap in. It then plays it back and allows you to change the tempo to suit your mood.

FIGURE 10.4 Wayne and Layne's Tactile Metronome follows the beat you set by tapping on the buzzer (wayneandlayne.com, $24.95). Credit: Wayne and Layne.

LushOne Synth

Iain Sharp builds complex modular synthesizers, like the LushOne shown in Figure 10.5. You can control it via a computer or musical keyboard, or even a joystick, variable resistor, or ultrasonic sensor. Modular synthesizers have multiple effects on separate parts of the circuit board (hence the name) and you can use patch cables to connect the various modules, giving you a ton of customizable effects.

FIGURE 10.5 Iain Sharp's LushOne synthesizer fits neatly inside this sweet treasure chest enclosure (lushprojects.com, $105). Credit: Iain Sharp.

ATARI PUNK CONSOLE

The Atari Punk Console was born in 1980, a project by Forrest Mims originally included in a Radio Shack booklet. It's a simple noisemaker driven by two 555 timer chips, one of which is set to output as an audio frequency oscillator, which creates a wave-shaped analog signal, whereas the other chip outputs as a monostable multivibrator—on and off—with both controlled by potentiometers. Together they create a fun variety of electronic noises.

The Vibrati Punk Console, shown in Figure 10.6, was created by Iain Sharp of Lushprojects.com. It's a variant of the Atari Punk Console that adds a low-frequency oscillator, which increases the noise and "dirtiness" of the sound.

FIGURE 10.6 The Vibrati Punk Console, shown here, is a variant of the APC.

Mini Project: Pushbutton Melody

Let's jump in and make some noise with your Arduino! Pushbutton Melody plays a song on a loudspeaker every time you press a button (see Figure 10.7). I programmed the song to play "Ode to Joy" and boy, does it sound electronic!

FIGURE 10.7 This simple project takes care of all of your electronic "Ode to Joy" needs.

PARTS LIST

This is a quick project so you only need a few things!

- A speaker. I suggest the 3", 8-ohm, 1-watt speaker from Adafruit: http://www.adafruit.com/products/1313
- A pushbutton (SparkFun P/N 97)
- A resistor; 220-ohm should do the trick
- Some jumpers and a breadboard

Instructions

This project is pretty easy to assemble. Just follow the wiring diagram shown in Figure 10.8 to see where to place the wires.

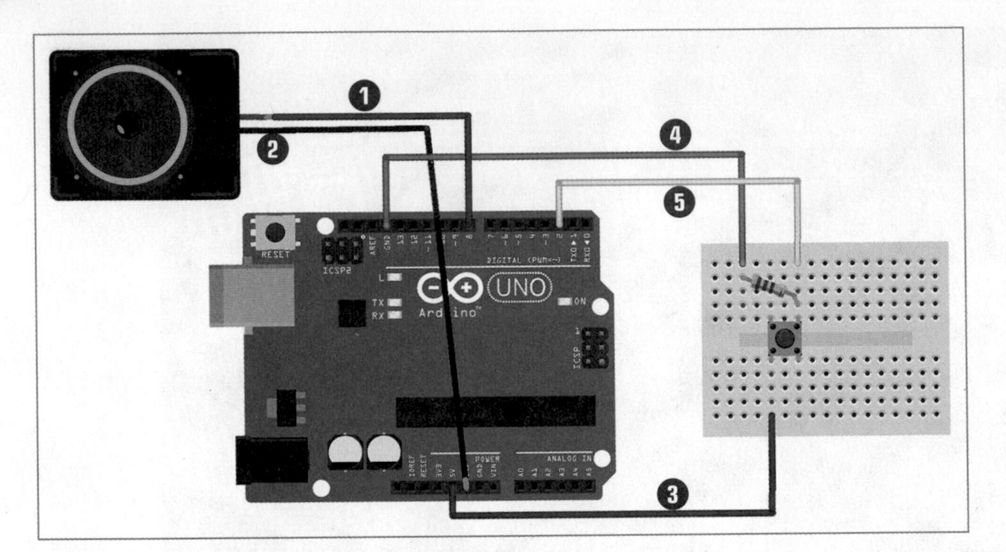

FIGURE 10.8 Wire up your Arduino as you see here.

1. Plug an 8-ohm speaker in to pin 8 (the red wire in Figure 10.8).
2. Plug the GND (black wire) from the speaker to the Arduino.
3. Connect the 5V port of the Arduino to one lead of the pushbutton. This is the green wire in Figure 10.8.
4. Connect the other lead of the pushbutton to GND, via the resistor (the blue wire).
5. Connect the pushbutton to pin 2 on the Arduino (yellow wire).

Pushbutton Melody Code

Although perhaps not very elegant, this code should get you started on learning how to make noise with your Arduino. Why do I say it's not elegant? Do you see the `Tone()` functions? I use a whole bunch of them rather than using a `For` loop, which I describe in Chapter 5, "Programming Arduino." In this case, I kept the sketch really obvious so you could mess around with it.

> **NOTE**
>
> **Code Available for Download**
>
> You don't have to enter all of this code by hand. Simply go to https://github.com/n1/Arduino-For-Beginners to download the free code.

```
const int buttonPin = 2;
const int ledPin =  13;
int pbState = 0;

void setup() {
  pinMode(buttonPin, INPUT);
}

void loop(){
  pbState = digitalRead(buttonPin);
  if (pbState == HIGH) {

tone(8, 247, 300);
delay(500);
tone(8, 247, 300);
delay(500);
tone(8, 262, 300);
delay(500);
tone(8, 294, 300);
delay(500);
tone(8, 294, 300);
delay(500);
tone(8, 262, 300);
delay(500);
tone(8, 247, 300);
delay(500);
tone(8, 220, 300);
delay(500);
tone(8, 196, 300);
delay(500);
tone(8, 196, 300);
delay(500);
tone(8, 220, 300);
delay(500);
tone(8, 247, 300);
delay(500);
tone(8, 247, 500);
delay(650);
tone(8, 220, 200);
delay(250);
tone(8, 220, 200);
delay(250);

  }
}
```

Project: Noisemaker

Having got our feet wet with the Pushbutton Melody, let's proceed to this chapter's full project: a Noisemaker that fits into your hand like a game controller (see Figure 10.9). It makes all sorts of cool noises, and you can build it yourself. Let's get started!

FIGURE 10.9 The Noisemaker project uses an Arduino to make crazy sounds.

PARTS LIST

Although the Noisemaker is a small build—barely larger than the Arduino inside it—it still takes a rather surprising diversity of parts to build. This is what you need:

- 1/8" plywood: I used two 3"×4" pieces to form the top and bottom.
- Speaker: I used a 1.5" model, 8-ohm, 0.25-watt. It fit nicely on the front of the Noisemaker, but it was too quiet for my tastes. I suggest SparkFun's 2", 8-ohm, 0.50-watt speaker, P/N 9151.
- Switch, Jameco P/N T100T1B1A1QN
- Two 2K potentiometers, Jameco P/N 31VA302-F3
- Light sensor, SparkFun P/N SEN-09088
- Resistors, I use 3.3K and 470-ohm resistors
- 9V battery
- Battery clip, Jameco P/N GBH-1009-R
- Battery power plug, Adafruit P/N 80
- Plastic standoffs, 3/8", #4-40; SparkFun P/N 10461
- Aluminum standoffs, I used hex #4-40, 1.5" male-female; Jameco P/N 166546
- Hot glue gun
- Drill and 11/64", 1/4", 3/8", and 3/4" bits
- 1" #4-40 bolts
- 1/4" #4-40 bolts
- An assortment of #4-40 nuts and washers
- Red, yellow, and black wire, Adafruit P/Ns 288, 289, and 290, respectively
- Heat-shrink tubing; SparkFun P/N 9353 offers a nice assortment

Instructions

Alright, let's get started with the actual build!

1. Cut the top and bottom out of 1/8" plywood, as shown in Figure 10.10. I made mine 3" × 4", and used a disk sander to round the corners.

FIGURE 10.10 The first step is to cut out the Noisemaker's top and bottom.

2. Starting with the bottom, drill out the holes for the standoffs and the battery clip using your 11/64" drill bit. Add the 1" #4-40 screws and plastic standoffs for the Arduino, as shown in Figure 10.11.

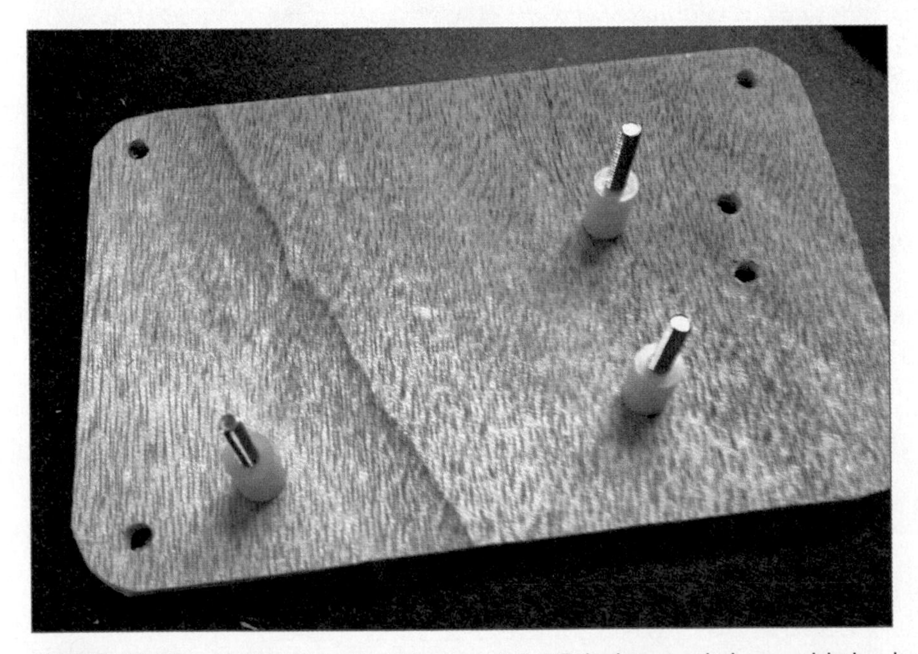

FIGURE 10.11 Drill the holes for the #4-40 bolts, and then add the bolts!

3. Attach the Arduino to the bottom plate, and add the battery clip using the 1/4" #4 screws. You can also attach the aluminum standoffs using some additional 1/4" screws. The bottom assembly should look like Figure 10.12.

FIGURE 10.12 Add the Arduino, battery pack, and standoffs.

4. On the top panel, drill out the holes for the speaker (3/4"), standoffs (11/64"), the potentiometers (3/8"), the light sensor (11/64"), and the switch (1/4"). It should look similar to what is shown in Figure 10.13.

FIGURE 10.13 The top panel awaits components.

5. Insert the back end of the speaker into the 3/4" hole and hot glue it in place, as shown in Figure 10.14. Make sure the wires are on the underside of the top panel, so they can reach the Arduino. Note that you might need a bigger speaker hole if you went with a different speaker than I did.

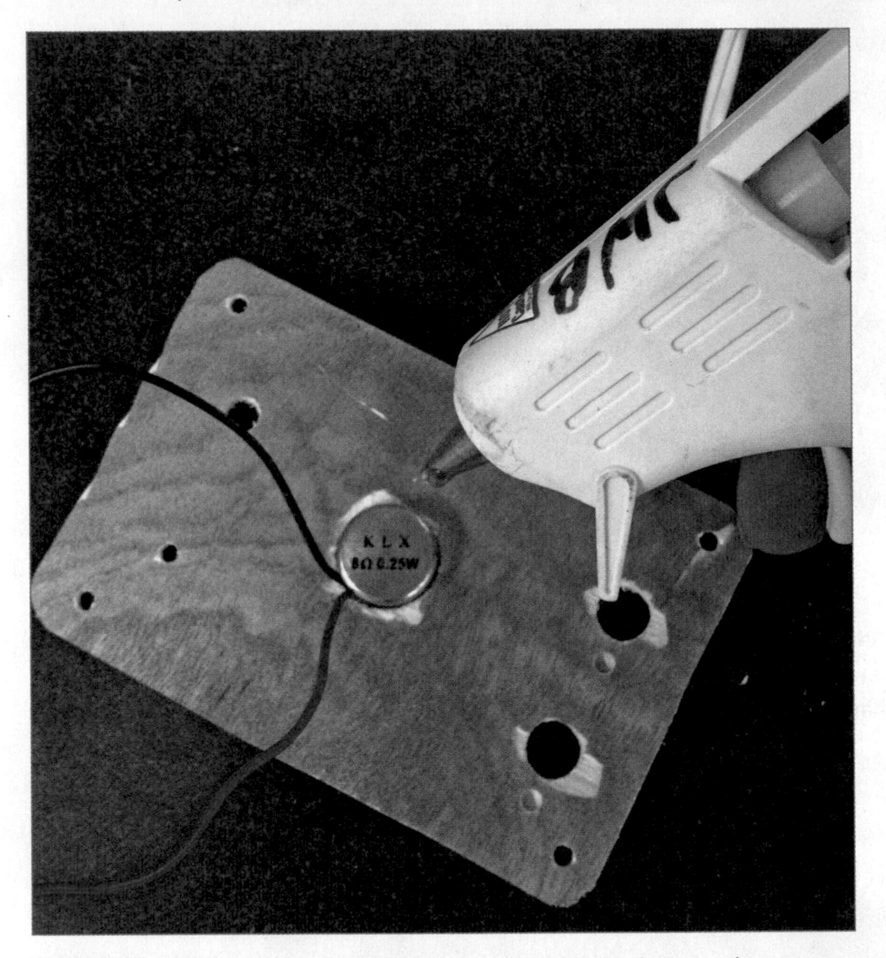

FIGURE 10.14 Hot glue the butt end of the speaker to the top panel.

6. Let's start wiring! Solder wires to the potentiometers as shown in Figure 10.15—you did this in Chapter 3, "How to Solder." Don't make the wires too long! About 5" of wire should do the trick. Do the same with the switch, and then attach the three components to the top panel.

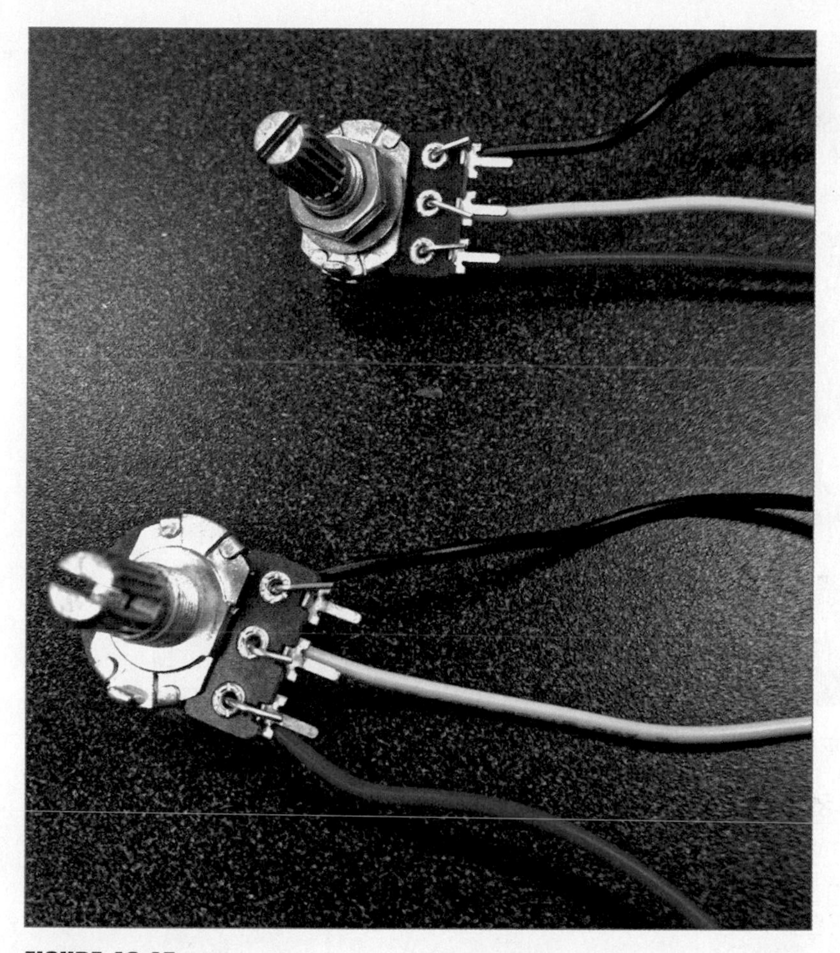

FIGURE 10.15 Wire up the potentiometers as you see here.

7. Insert the light sensor from the top and hot glue it in place (see Figure 10.16) from the bottom, making sure you don't goop up the leads!

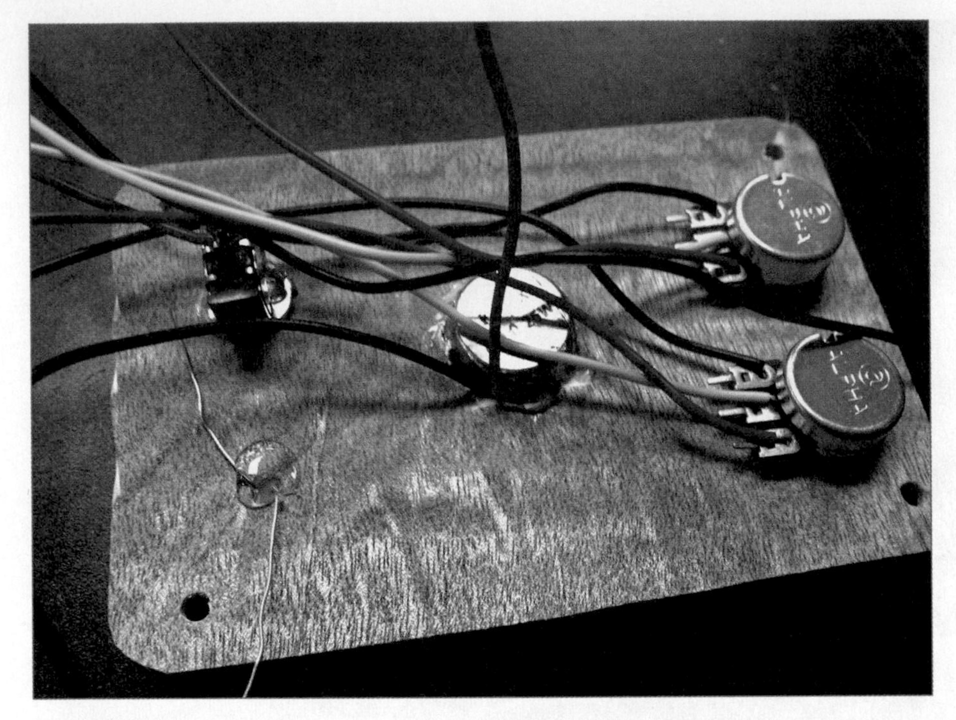

FIGURE 10.16 Hot glue the photo resistor to the top panel.

8. Wire up a 3.3K-ohm resistor to the light sensor's ground wire, then solder in a yellow wire for data. The tail end of the resistor should have a black wire soldered onto it, and this becomes the ground. You can see this in Figure 10.17. It gets plugged into a GND pin of the Arduino.

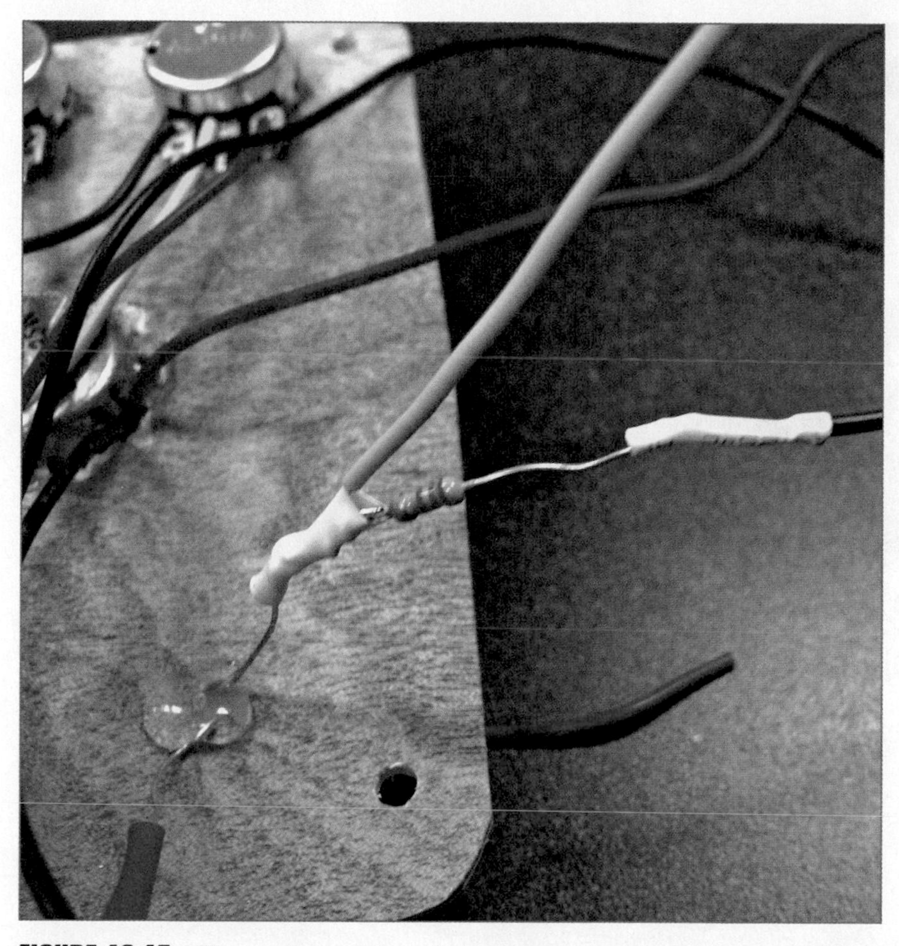

FIGURE 10.17 The ground wire of the photo resistor gets a resistor.

9. Repeat step 8, but with the switch. The ground wire of the switch gets a 470-ohm resistor and a second wire, along with a length of wire soldered onto the end of the resistor. The end with the resistor becomes the ground, while the wire without the resistor becomes data.

   You can see these wires in Figure 10.18. Plug the ground wire into a free GND pin on the Arduino.

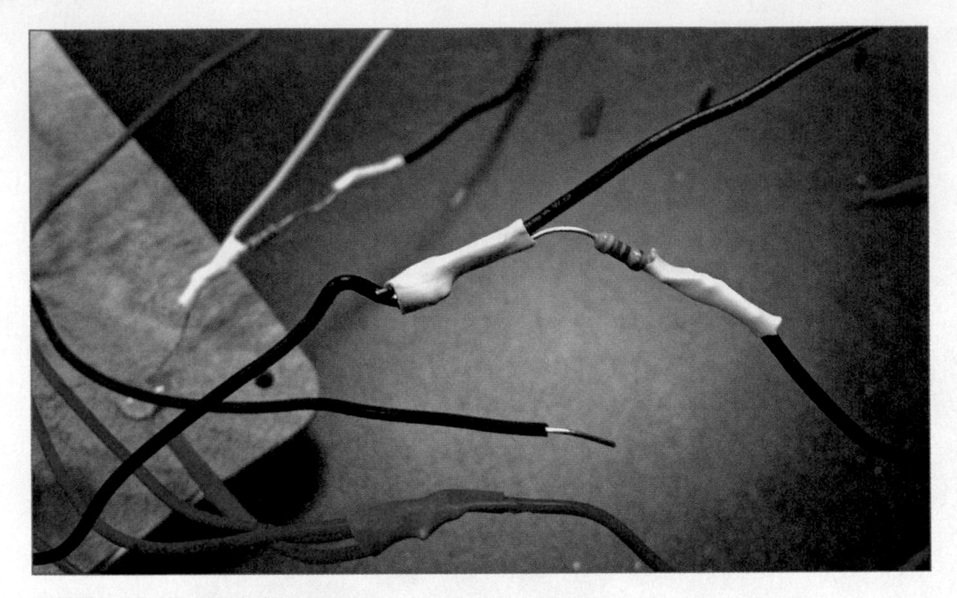

FIGURE 10.18 The switch also gets a resistor.

10. Solder the black (ground) wires of the potentiometers to the ground wire of the speaker, solder in a length of wire, and then cover in heat-shrink tubing. You're basically combining the three ground wires into one, as shown in Figure 10.19. Plug this into a GND pin of the Arduino.

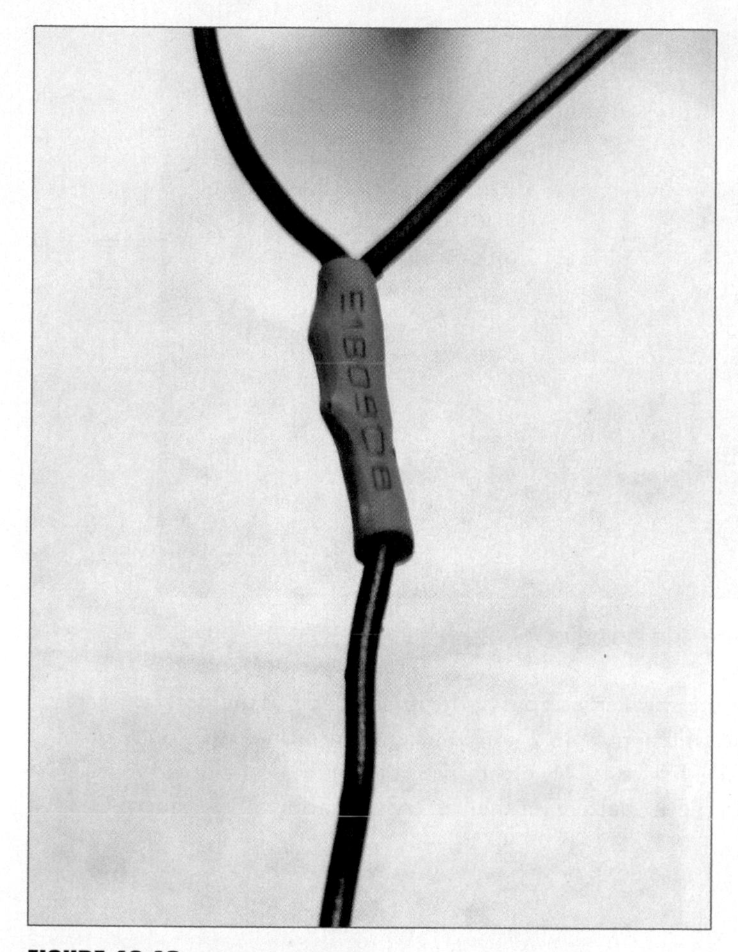

FIGURE 10.19 Combine the grounds into one.

11. Solder the positive wires of the potentiometers, light sensor, and switch together, combining the four wires into one as you did in step 8. Cover the join with heat-shrink tubing. It should look like Figure 10.20! This gets plugged into the 5V pin of the Arduino.

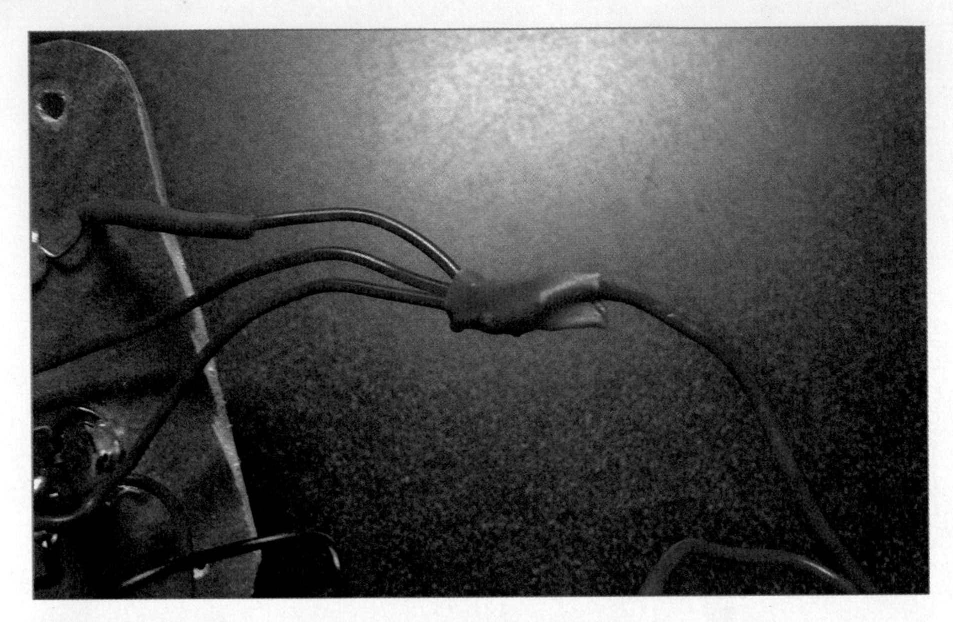

FIGURE 10.20 Now combine the positive leads.

12. Plug the positive wire of the speaker into pin 8, the middle lead of the potentiometers into pins A1 and A2, the switch into pin 2, and the light sensor into A0. You're done! The circuit should look like Figure 10.21, except possibly not having a breadboard. All you have to do is plug in the 9V battery via the battery plug and you're golden!

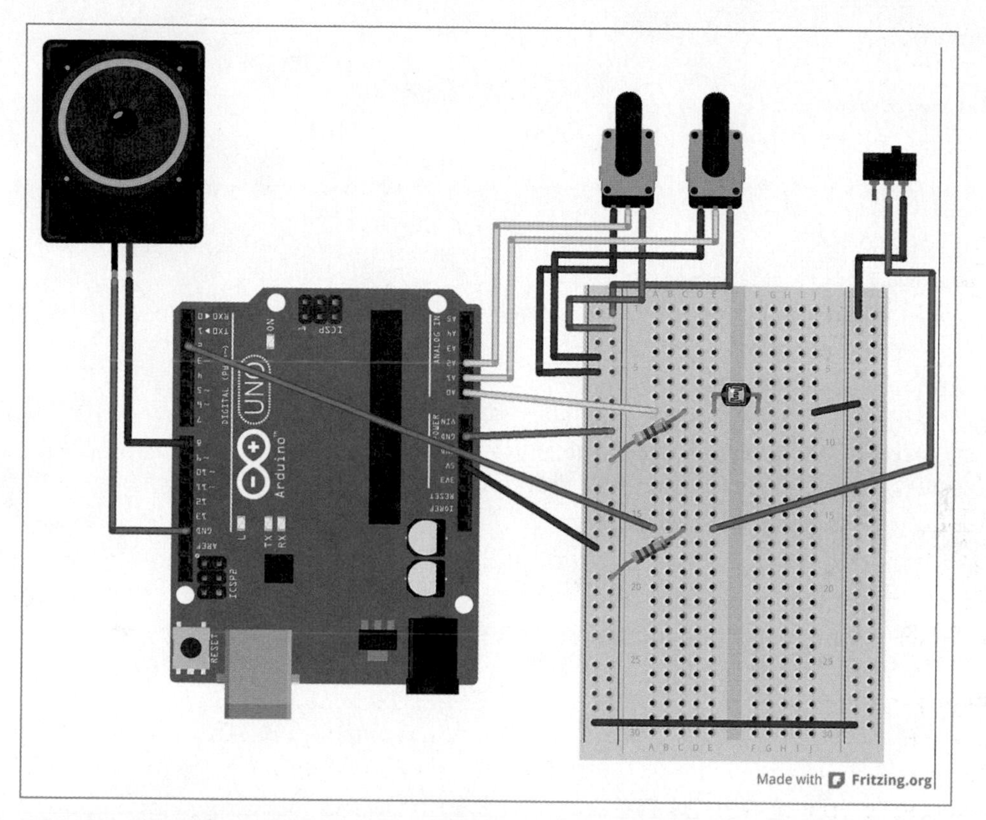

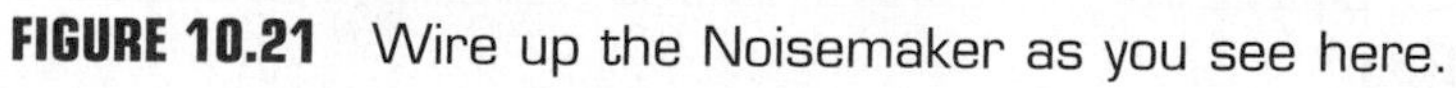
FIGURE 10.21 Wire up the Noisemaker as you see here.

Noisemaker Code

The Noisemaker code is elegantly simple.

> **NOTE**
>
> **Code Available for Download**
>
> You don't have to enter all of this code by hand. Simply go to https://github.com/n1/Arduino-For-Beginners to download the free code.

```
void setup()
{
  Serial.begin(9600);
}

void loop()
{
  int sensorReading = analogRead(A0);
  int pot1 = analogRead(A1);
  int pot2 = analogRead(A2);
  int switch1 = digitalRead(2);

  int thisPitch = map(sensorReading, 600, 1000, 1000, 100);
  int potDelay = map(pot1, 0, 1023, 1, 100);
  int potDur = map(pot2, 0, 1023, 1, 50);


  if (switch1 == HIGH) {
    tone(8, thisPitch, potDur);
    delay(potDelay);
  }
}
```

The Next Chapter

In Chapter 11, "Measuring Time," you learn how Arduino marks the passage of hours and minutes and how to help it do a better job of accurate timekeeping. Then you'll build some indoor wind chimes that ring on the hour, rather than relying on nature to do the work.

11

Measuring Time

How exactly does a robot tell time—perhaps it looks at a clock like the rest of us? That sounds flip, but it's actually true: It's possible to have the Arduino look up an Internet "time server" and get the official time. More prosaically, you can also have the Arduino use its (not terribly accurate) internal timer to tell time, or use a dedicated real-time clock module (RTC) to keep track of hours and minutes. In this chapter, we explore a variety of ways in which an Arduino can keep track of time, and then you'll tackle the project for this chapter, a motor-controlled "wind chime" that triggers on the hour. Figure 11.1 shows an example of an interactive Arduino-based clock.

FIGURE 11.1 Nootropic Designs' Defusable Clock is an interactive Arduino-based clock that looks like a Hollywood bomb! Credit: Nootropic Design

Time Server

One way for your project to keep track of time is to continuously access an Internet-based time server via Wi-Fi, usually using a Wi-Fi shield, as shown in Figure 11.2. These sites, called NTP (network time protocol) servers, are resources providing accurate time to Internet-connected gadgets. If you have a smartphone, you probably have noticed it never needs to be set, automatically knowing the correct time. An NTP server gets the credit!

> **NOTE**
>
> **Learn More About Accessing an NTP Server**
>
> For a tutorial on how to access an NTP server with your Arduino and Wi-Fi shield, see this page on the Arduino site: http://arduino.cc/en/Tutorial/UdpNTPClient.

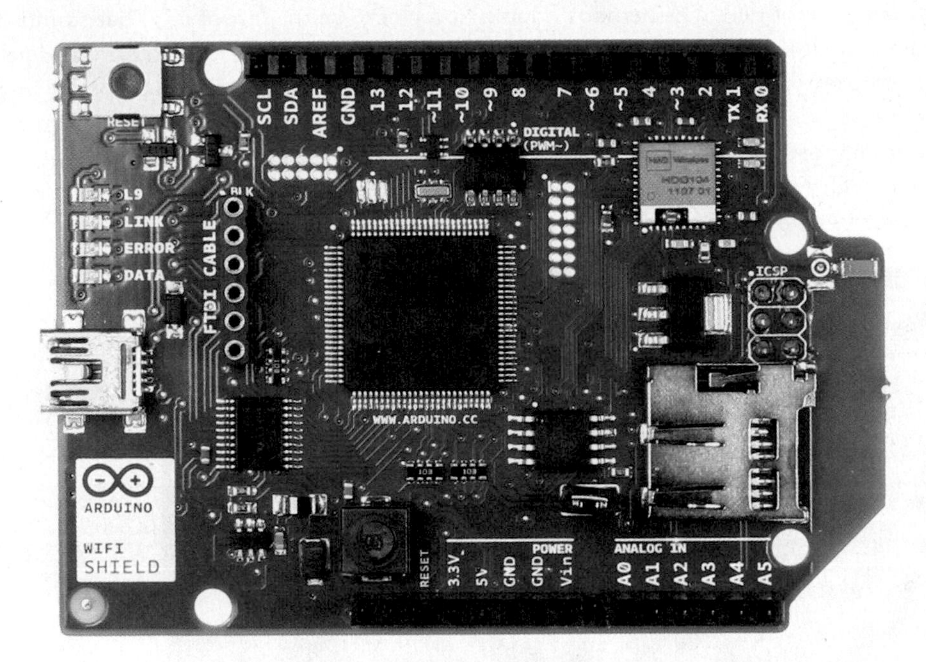

FIGURE 11.2 Arduino's Wi-Fi shield gives your Arduino robot the ability to connect to wireless networks. Credit: Arduino.cc

Arduino's Timer

The Arduino's main chip, the ATmega328P (see Figure 11.3), contains a timing circuit that does a fairly okay job at keeping time. Just as you would use the command `delay(1000);` to tell the Arduino to wait 1,000 milliseconds, the timer built in to the ATmega tells it when that time has passed.

FIGURE 11.3 The Arduino's microcontroller chip also has a timing function that you can harness to keep track of time.

Due to the board's modest architecture, it can keep track of time for only 49 days before it runs out of memory and must reset. In addition, accuracy is not precise. Sticklers for precision will be upset to learn that the ATmega loses about two seconds per day. Basically, after it reaches that 49-day mark it will already be wildly inaccurate, around 100 seconds off the mark. Most tinkerers use an RTC if they want accurate measurement of time.

Real-Time Clock (RTC) Module

Another option for keeping track of time is to connect a real-time clock (RTC) module like the ChronoDot shown in Figure 11.4. An RTC consists of a circuit board with a highly accurate timer chip, as well as a coin-cell battery backup that keeps the time set even if the board is unplugged. When properly configured, the ChronoDot loses less than a minute per year thanks to its temperature-controlled switch, and a fresh battery will keep the time for around eight years.

FIGURE 11.4 A ChronoDot RTC module plugged into a breadboard.

Mini Project: Digital Clock

For this mini project, you'll make a digital clock (see Figure 11.5) that keeps perfect time thanks to a real-time clock module, a small board that has a timer chip and battery backup so that it never forgets the time. It's not pretty, but you could definitely put it in some sort of decorative case.

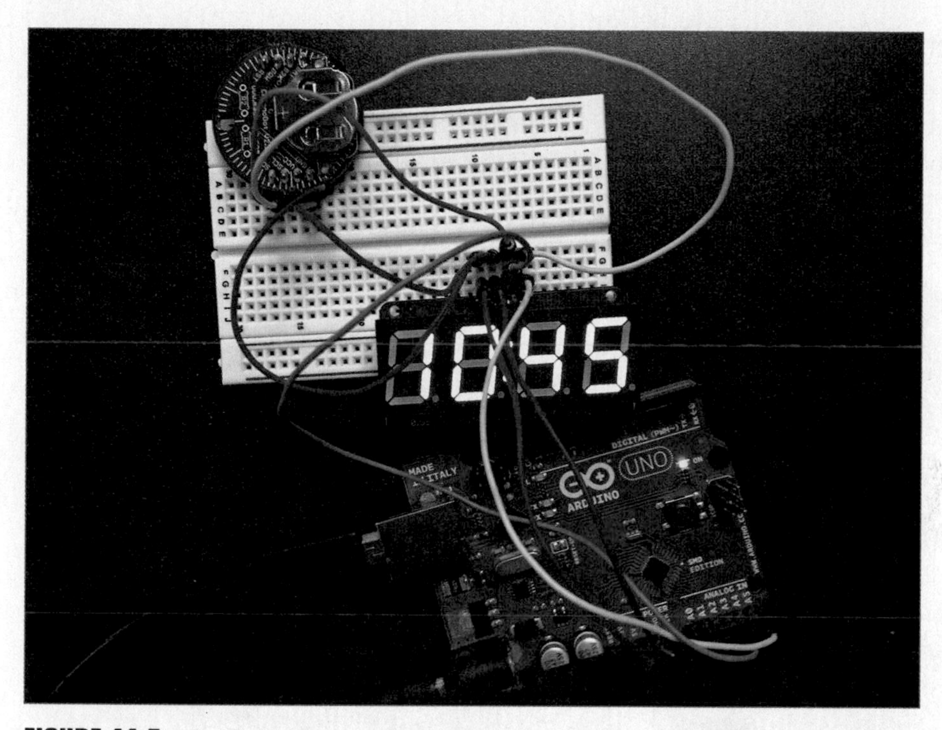

FIGURE 11.5 An Arduino, seven-segment display, and RTC module are all you need to create a clock!

PARTS LIST

You'll need just a few things to make the digital clock:

- Arduino Uno with power supply
- RTC module: I used the ChronoDot RTC (Adafruit P/N 255), but you can also use the cheaper DS1307 RTC breakout board kit (Adafruit P/N 264).
- Adafruit Seven-Segment Backpack: This invaluable board (P/N 878) consists of a seven-segment display with a circuit board designed to make it easier to bread board.
- Half-size breadboard: Adafruit P/N 64
- Jumpers

Instructions

Let's wire up the digital clock, following along with Figure 11.6. Note that the image I used for the RTC is the DS1307 I mentioned earlier in this chapter. Functionally it works the same as the ChronoDot, and they both use the same Arduino library, so for the purposes of this project, which one you use doesn't matter too much. Let's get started!

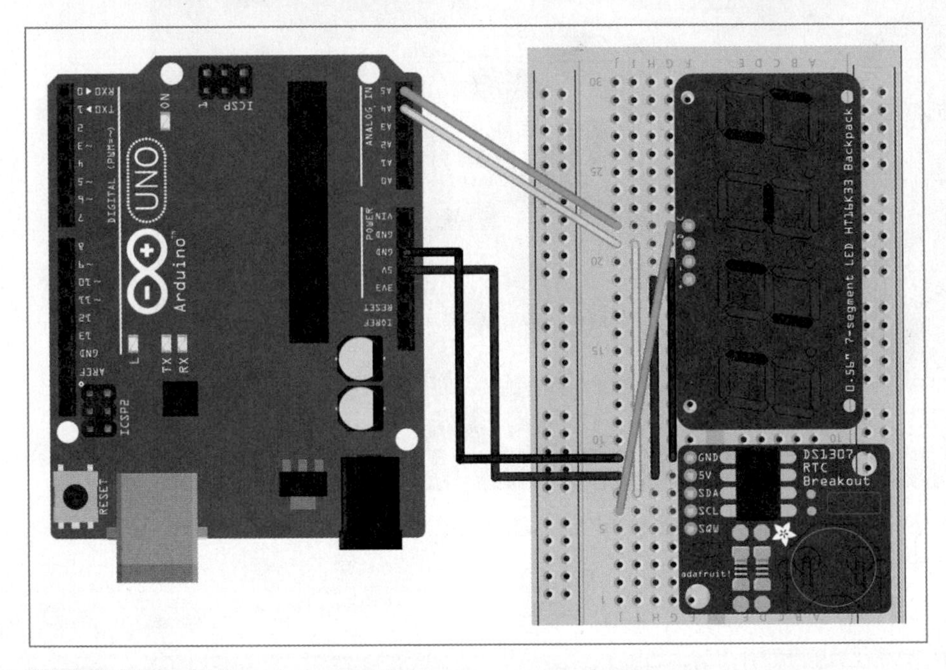

FIGURE 11.6 Wire up your clock as you see here.

1. Plug in the seven-segment backpack to the breadboard, making sure to leave plenty of room on those rows for jumpers.
2. Attach the RTC module. This should also leave room for jumpers, as shown in Figure 11.6.
3. Wire up the boards. This is a little tricky because both the seven-segment backpack and the RTC share the same four pins on the Arduino!
    1. Connect the "+" pin on the backpack to the 5V pin on the RTC (marked as "VCC" on the ChronoDot) and to the 5V pin on the Arduino. This is the red wire in Figure 11.6.
    2. Connect the "–" (ground) pin on the backpack to the GND pin on the RTC, and then to a GND pin on the Arduino. This is the black wire in Figure 11.6.
    3. Connect the "C" (clock) pin on the backpack to the SCL pin on the RTC and then to pin A5 (that's analog, not digital!) on the Arduino. This is the green wire in Figure 11.6.
    4. Connect the "D" (data) pin on the backpack to the SDA pin on the RTC and pin A4 on the Arduino. This is the yellow wire in Figure 11.6.

You're finished with hardware! Now, let's program the Arduino.

Digital Clock Code

Now you can upload the sketch to the Arduino. As with previous chapters, you can download this sketch from https://github.com/n1/Arduino-For-Beginners.

```
// This code is based off of Adafruit's example text for the RTC.

#include <Wire.h>
#include "Adafruit_LEDBackpack.h"
#include "Adafruit_GFX.h"
#include "RTClib.h"

RTC_DS1307 RTC;
Adafruit_7segment disp = Adafruit_7segment();

void setup()
{
  Wire.begin();
  RTC.begin();
  if (! RTC.isrunning())
  {
    RTC.adjust(DateTime(__DATE__, __TIME__));
  }
  disp.begin(0x70);
}

void loop()
{
  disp.print(getDecimalTime());
  disp.drawColon(true);
  disp.writeDisplay();
  delay(500);
}

int getDecimalTime()
{
  DateTime now = RTC.now();
  int decimalTime = now.hour() * 100 + now.minute();
  return decimalTime;
}
```

Project: Indoor Wind Chime

For this chapter's project you're going to build a sweet wind chime (see Figure 11.7) that relies on its real-time clock module to tell it when to chime. You'll also build a geometric enclosure to house the electronics, using a tool called a CNC router.

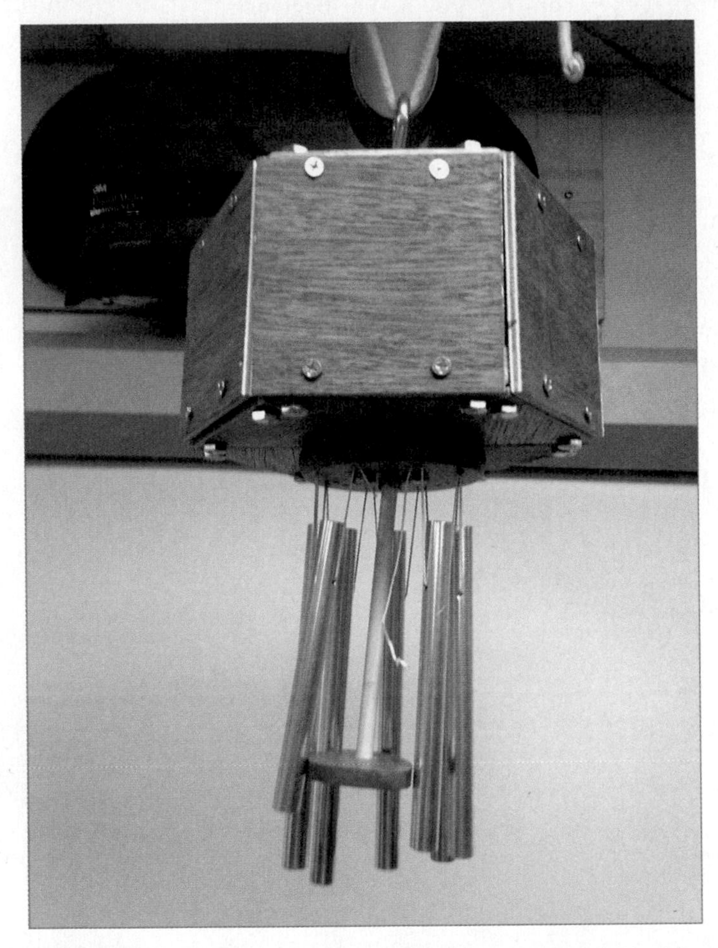

FIGURE 11.7 Learn how to build this sweet wind chime!

PARTS LIST

- Arduino
- Servo (I used a HiTec HS-322HD servo, Jameco P/N 33322.)
- Servo horns (See the following section; a number of horns come with the HiTec; these should be fine.)
- ChronoDot RTC module
- Mini breadboard
- 9V battery and battery clip (Digi-Key P/N BC22AAW-ND)
- 1/4" dowel (you'll need about 8" to a foot)
- Wind chime (I used a Gregorian Chimes Soprano wind chime, SKU 28375-00651.)
- 5mm plywood for the enclosure
- 1" pine board for the support blocks
- Eye bolt and nut (The Home Depot P/N 217445)
- #8 × 1/2" wood screws
- #6 × 2" wood screws
- #4 × 1/2" wood screws
- 24 1/4" × 1 1/2" bolts with locking washers and nuts
- 12 1/4" × 1" bolts with locking washers and nuts
- Drill press and a variety of drill bits
- Chop saw
- Table saw
- Hole saw
- Belt sander

Servo Horns

Servos connect to wheels, axles, and other parts of a robot using connectors called servo horns. These consist of a bars or discs (see Figure 11.8) studded with screw holes, and featuring a toothed lug that fits over the servo's hub. A screw secures the horn.

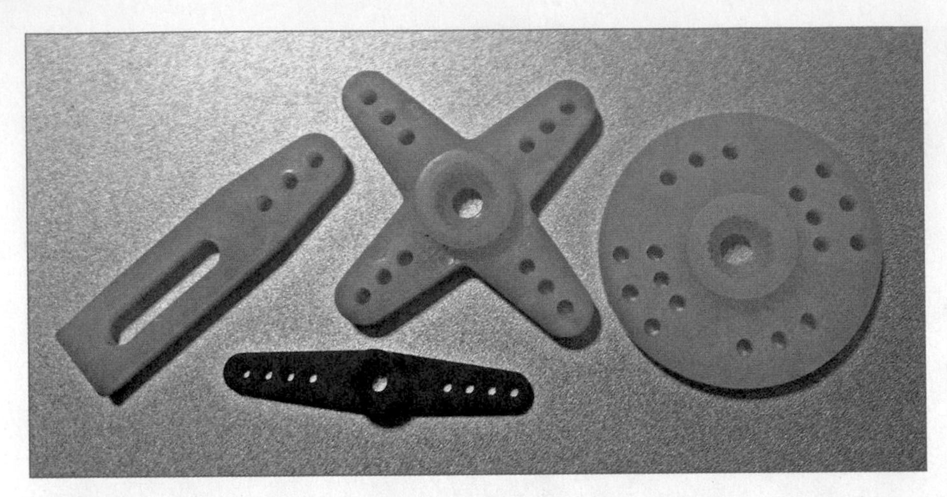

FIGURE 11.8 Servo horns help connect your robot to the servos that move it.

Instructions

Follow these steps to build your indoor wind chimes:

1. Mill the triangle shapes used to make the enclosure. I used a CNC router, as shown in Figure 11.9. If you're making your own, they're 3" equilateral triangles with 1/4" holes drilled in the three corners. You can download the .DXF files I used to mill the triangles from https://github.com/n1/Arduino-For-Beginners.

FIGURE 11.9 A CNC router cuts out the shapes you need for this project.

2. After you've cut out the triangles, sand them down on a belt sander (see Figure 11.10) so their edges are smooth.

FIGURE 11.10 Smooth down the burrs on a belt sander.

3. Cut support blocks out of the 1" pine. Make these look like the one in Figure 11.11. These blocks reinforce the top and allow you to attach the sides. Make 12 total because you'll need some for the bottom.

FIGURE 11.11 The support blocks secure the various triangle-shaped pieces that make up the top and bottom.

4. Cut a reinforcing disc out of the 5mm plywood using the hole saw; it should look just like the one in Figure 11.12.

FIGURE 11.12 This reinforcing disc keeps the top of the enclosure in order.

5. Drill six 1/4" holes surrounding a 5/8" hole. Make sure to align the six surrounding holes so that they fit with the six holes at the center of the top of the enclosure.
6. Assemble the top, using the 1/4" × 1 1/2" bolts to attach the triangle pieces to the support blocks, then add the disc and eye bolt in the middle. It should look just like Figure 11.13. You'll probably have to redrill the center hole because the points of the triangles get in the way.

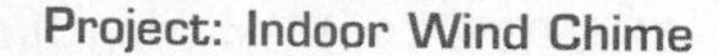

FIGURE 11.13 Make a hexagonal enclosure for your indoor wind chime!

7. Build the bottom (see Figure 11.14) like the top, with a couple of differences: One is that it doesn't get an eye bolt. Instead, a dowel will protrude from the center bottom of the enclosure. Another is that instead of a disc in the center, you'll simply use the disc from the wind chime. Screw seven holes in it just as you did with the top disk. (Note in Figure 11.14, I show only three of the holes being populated with bolts—I just ran out of bolts!)

FIGURE 11.14 Drill seven holes in the top portion of the wind chime and attach it to the bottom.

8. Cut a 2" disc out of the 5mm plywood using your hole saw, and give it a 1/4" hole in the center. Screw the servo horn to the disc using the #4 × 1/2" wood screws, then glue the dowel to the hole in the disc. It should look like Figure 11.15. Like the disc in the figure, it doesn't have to be beautiful—it's just a convenient way to attach the servo horn to the dowel.

FIGURE 11.15 It doesn't have to be pretty!

9. Cut side panels out of the 5mm plywood. They should be 4" on a side (see Figure 11.16). You'll need to drill holes in the wood to attach the panels to the top and bottom support blocks; placement of these holes isn't super tricky, as long as it looks good and connects to the support blocks. If you want to, at this time you can attach them to the top of the enclosure.

FIGURE 11.16 The side panels, ready for installation.

10. Install the motor in the top panel, using strips of wood cut from the 5mm plywood, with 1" spacer blocks cut from the pine. Use the #8 × 1/2" screws to attach the servo to the strips, then use the #6 × 2" screws to attach the strips to the support blocks. Really, the only considerations are that the strips of wood are high enough so that the servo doesn't bump into the top, and that the servo's hub is aligned with the dowel-hole in the bottom of the enclosure. You can see how it should look in Figure 11.17.

FIGURE 11.17 Looking into the enclosure from the bottom, you see the servo mounted in the center.

11. Attach the Arduino, breadboard, and battery pack to the inside of the enclosure. I suggest bolting the Arduino to a side panel with #4 × 1" machine screws and hot gluing the breadboard and battery pack. It should look more or less like you see in Figure 11.18.

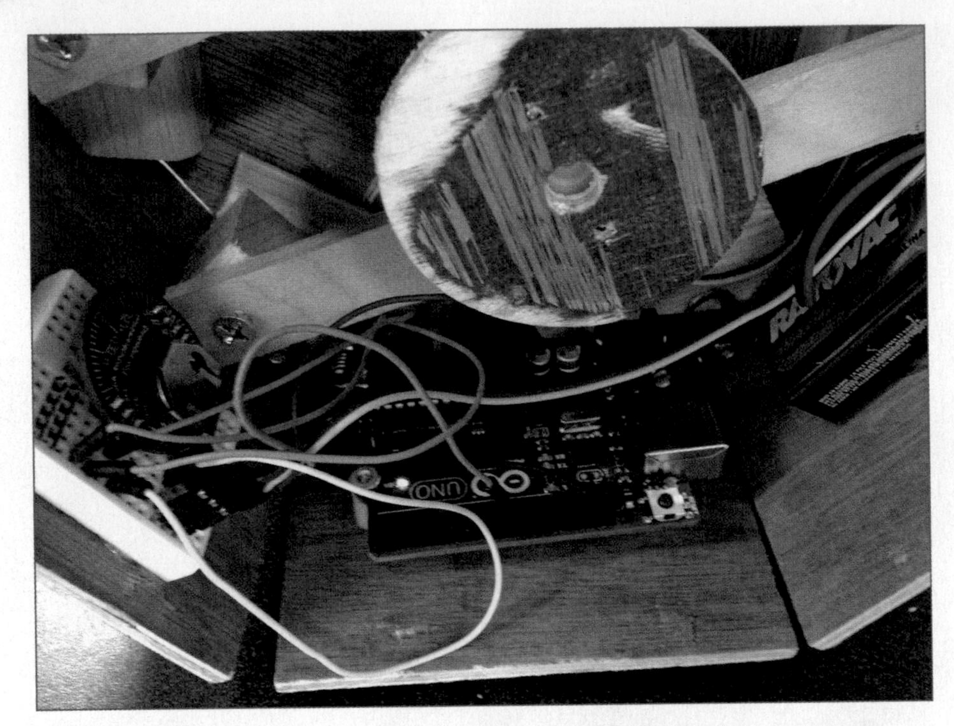

FIGURE 11.18 Attach the electronics to the inside of the enclosure.

12. Wire up the various components, as shown in Figure 11.19.

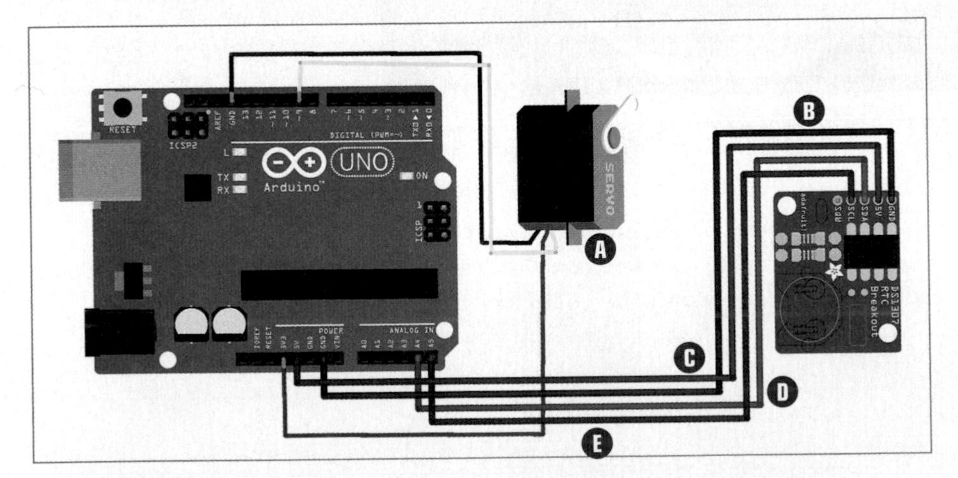

FIGURE 11.19 Connect the various components as you see here.

A Connect the servo. The yellow wire goes to pin 9 of the Arduino, the red wire goes to the 3.3V pin of the Arduino, and the black wire connects to one of the Arduino's GND pins.

B Connect the GND pin of the RTC to a GND pin on the Arduino.

C Connect the 5V pin of the RTC to the 5V pin on the Arduino.

D Connect the SDA (data) pin of the RTC to A4 on the Arduino.

E Connect the SCL (clock) pin of the RTC to A5 on the Arduino.

13. Drill a hole in the clapper that came with the wind chimes (see Figure 11.20). The clapper will rotate with the servo and bang on the chimes to make noise. However, don't glue the dowel in the clapper's hole just yet!

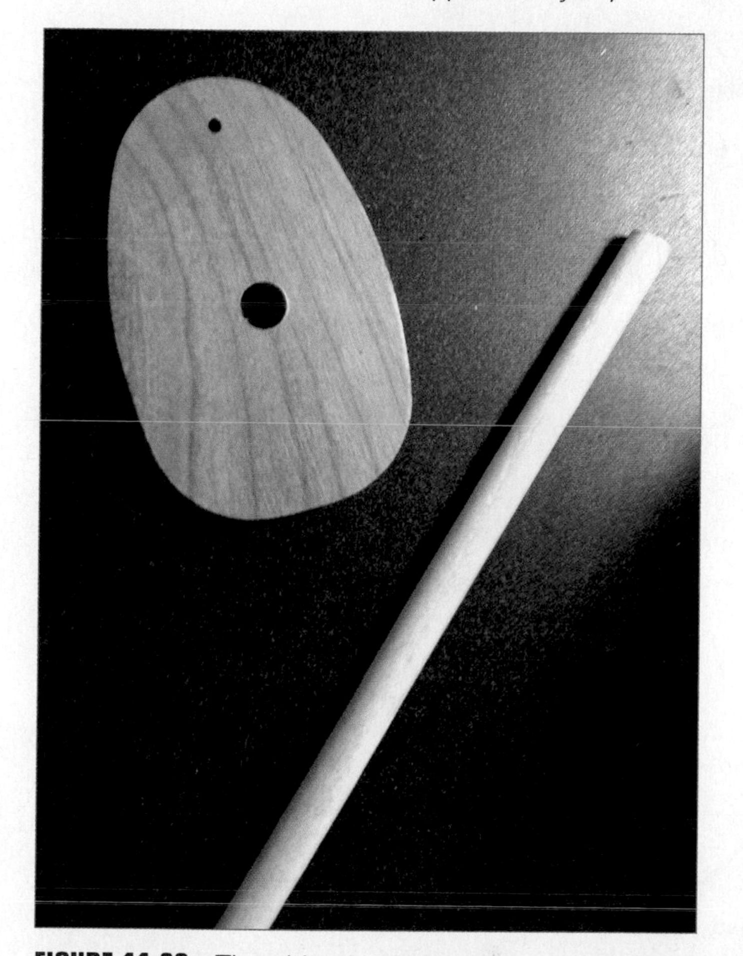

FIGURE 11.20 The chime's clapper gets repurposed as the chime's clapper.

14. Finish up by screwing the side panels onto the bottom hexagon. As a final step, glue the clapper onto the end of the dowel (see Figure 11.21).

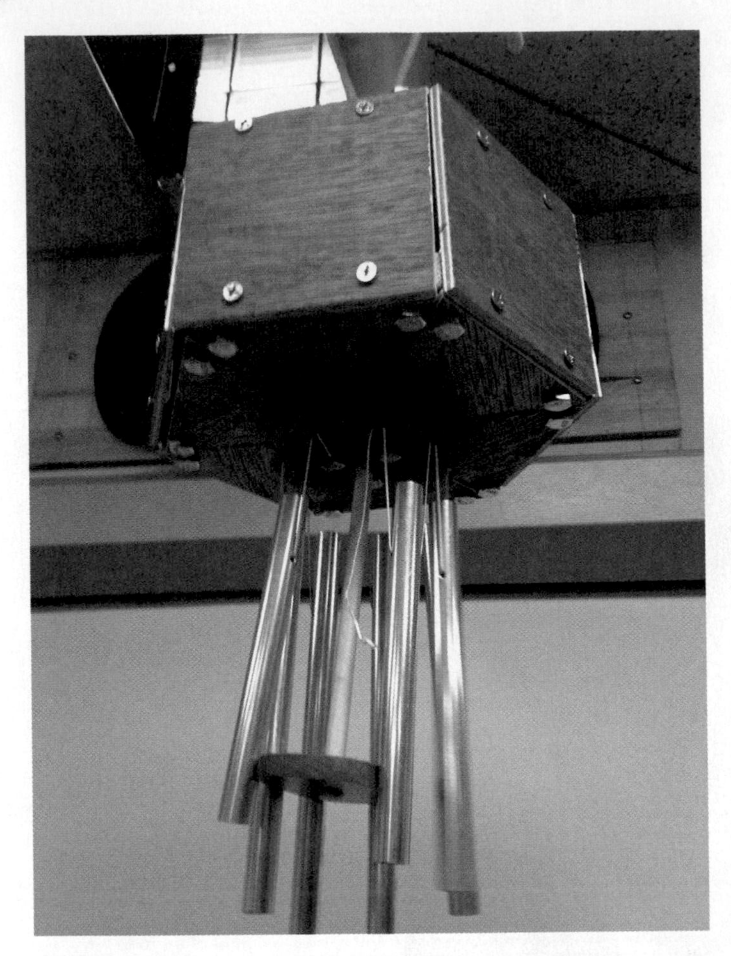

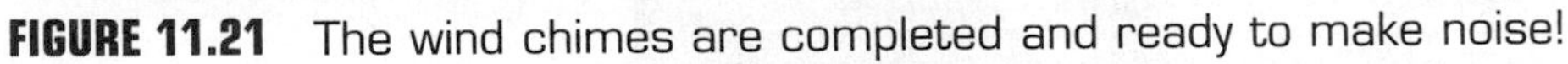

FIGURE 11.21 The wind chimes are completed and ready to make noise!

Code

The code is fairly simple, consisting of just a function to pull the time off the RTC module and another to rotate the servo when the minutes read as zero.

> **NOTE**
>
> **Code Available for Download**
>
> You don't have to enter all of this code by hand. Simply go to https://github.com/n1/Arduino-For-Beginners to download the free code.

```
#include <Wire.h>
#include "RTClib.h"
#include <Servo.h>

Servo myservo;
RTC_Millis RTC;
  int pos = 0;


void setup()
{


      Serial.begin(57600);

  RTC.begin(DateTime(__DATE__, __TIME__));

  myservo.attach(9);  // attaches the servo on pin 9 to the servo object

}

void loop() {



      DateTime now = RTC.now();

      int decimalTime = now.hour() * 100 + now.minute();
      Serial.print(decimalTime);
delay (60000);

    Serial.println();


    if (now.minute() == 0) {

  for(pos = 0; pos < 180; pos += 1)
  {
    myservo.write(pos);
    delay(15);
  }
  }


}
```

Computer Numerically Controlled (CNC) Tools

CNC tools, like the CNC router I used to create the wooden panels used in this project's enclosure, take direction from the computer to move a milling tool around, grinding, drilling, and shaping pieces of wood, metal, and other materials. Figure 11.22 shows a CNC router.

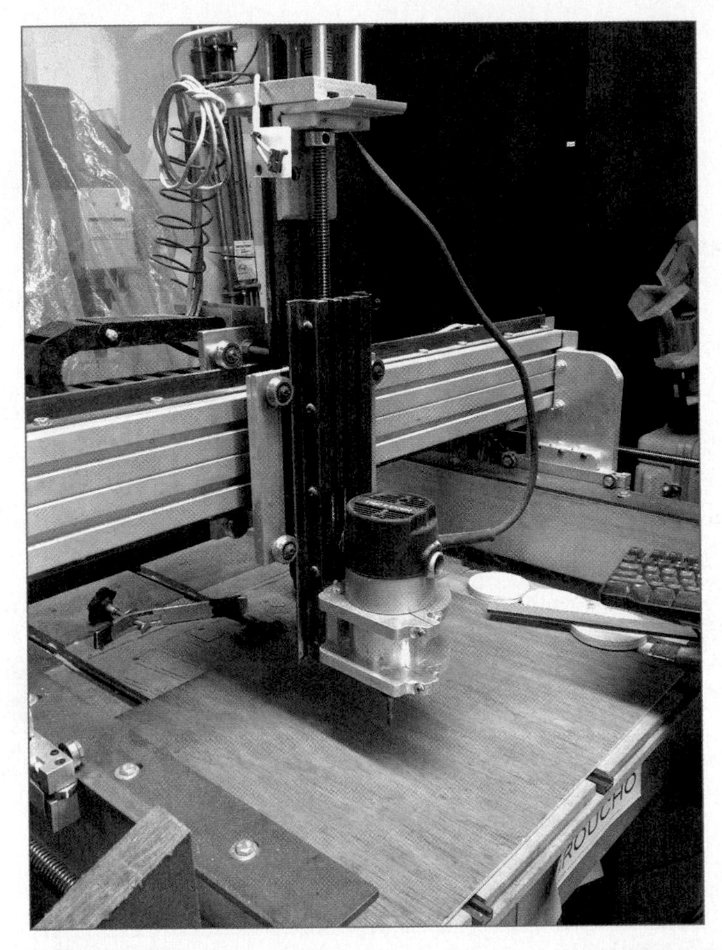

FIGURE 11.22 The CNC router is a valuable tool for precision cutting of wood and metal.

This is how the CNC router works:

1. Design whatever it is you want cut out, usually using a vector art program such as Adobe Illustrator or CorelDRAW.
2. Use a CNC utility to get the art ready to mill. One example of this is Vectric Cut2D (vectric.com), which guides you through the process of deciding how each element will

be milled, and in what order. For instance, say you have a 1" circle in your design. In Cut2D, you can tell the router to move in a circle to cut out that shape.

NOTE

Carving a 3D Shape Is Possible

Another factor to keep in mind is that the CNC mill can cut into a thick block of material, essentially carving a 3D shape with its tools.

3. Place the material on the CNC router's bed and clamp it down, to ensure that the material doesn't move as the router bit shapes it. Similarly, make sure you leave plenty of material around the shape you're cutting so it's supported throughout the cutting process.
4. Load up the design file on the CNC router's workstation. You might want to do a "dry run" of your job: This is like running through it with the tool a few inches above the material so you can see everywhere it goes. If the tool appears to move beyond the edge of the material, or bumps into a clamp, you know to fix the job before starting milling!
5. Finally, running a CNC router can be somewhat hazardous, with chips of material flying off the machine—wear goggles! It can also be noisy, so make sure you wear ear protection as well.

The Next Chapter

In Chapter 12, "Safely Working with High Voltage," you learn to harness the power of the outlet—safely!—and how to create an Arduino-controlled lava lamp for your home.

12

Safely Working with High Voltage

Zap! We've successfully trained ourselves to fear high voltage electricity, and rightfully so! It's hard to kill yourself with an AA cell, but sticking your tongue in an outlet is sure disaster. In this chapter, you'll explore a couple of ways to safely use high voltage in your projects. You'll then make an Arduino-controlled Lava Lamp Buddy that turns your lava lamp on and off on a schedule, or at the command of a remote control (see Figure 12.1). It's just what every lava lamp needs!

FIGURE 12.1 The Lava Lamp Buddy controls your favorite bubbling light fixture.

Lesson: Controlling High Voltage

The secret to controlling high voltage is to not have anything to do with it! I joke, but that's actually pretty good advice. Instead, let's allow a clever electronic component called a relay (see Figure 12.2) do the dangerous work. I mentioned relays in Chapter 1, "Arduino Cram Session." They're essentially switches that an Arduino can trigger; the relay handles the voltage so you never need to mess with it. Of course, relays are just electronic components and need a framework, such as a circuit board, to operate within. The following sections detail three products that feature relays and that you can use to work with high voltage.

FIGURE 12.2 A relay is the ticket to controlling high voltage.

PowerSwitch Tail

A PowerSwitch Tail (Adafruit P/N 268) looks like a short extension cord with a power supply built in, as shown in Figure 12.3. That's basically what it is, except that the power supply brick has ports for adding wires, allowing you to trigger the voltage with a single wire from an Arduino pin. The PowerSwitch Tail also includes a ground port.

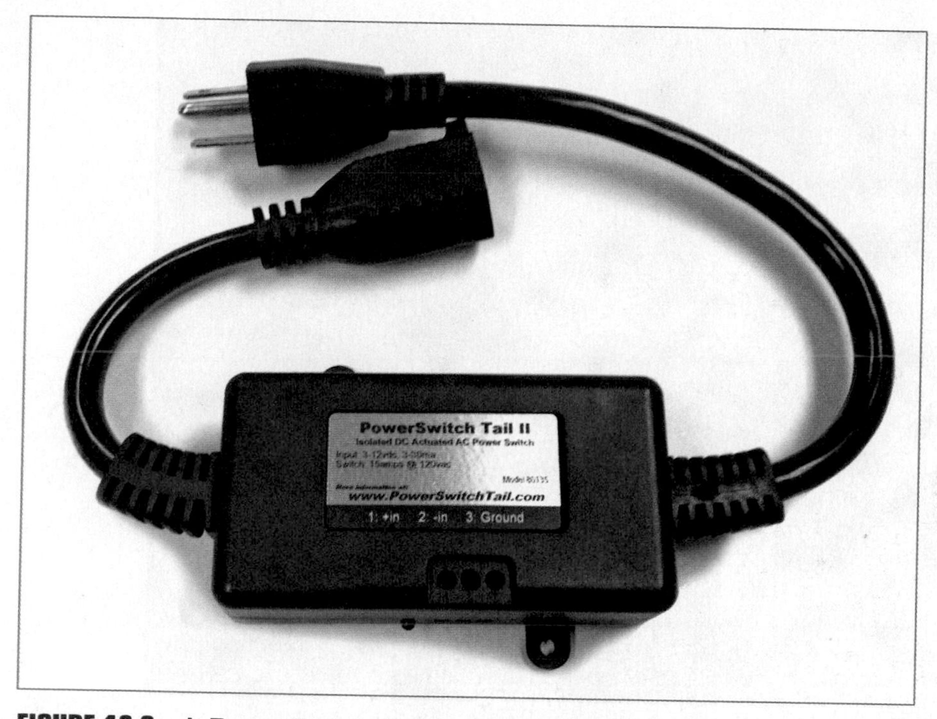

FIGURE 12.3 A PowerSwitch Tail is essentially a short power cord with a relay board built in.

What sets the PowerSwitch Tail apart from the competition is that it really is foolproof. Can you plug an appliance into an outlet? Then you can work a PowerSwitch Tail.

EMSL Simple Relay Shield

Offering a completely different configuration than the PowerSwitch Tail, the Simple Relay Shield (EMSL P/N 544; see Figure 12.4) created by Bay Area hardware hackers Evil Mad Scientist Laboratories, or EMSL for short (evilmadscientist.com), works as an Arduino shield, meaning that it's a circuit board with pins on the bottom, allowing it to be inserted into headers on the Arduino. The shield itself also has headers, allowing you to not only control the relay but also monitor sensors or light up LEDs as you normally would.

FIGURE 12.4 The Simple Relay Shield adds a relay to your Arduino. You can see the edges of the Arduino under the Simple Relay Shield shown here.

The setup has a couple of downsides:

- You can't use it to handle standard 110V current, which you can with the PowerSwitch Tail. Its maximum voltage is 40V/5A AC or 24V/5A DC.
- You have to connect the high-voltage wires to the shield manually, which means potentially exposing yourself to nasty shocks.

On the upside, it costs less than half as much as a PowerSwitch Tail!

Beefcake Relay Control Board

SparkFun's Beefcake Relay Control Board (P/N 11042) is inexpensive—$8—and easy to use. The Beefcake also has a monster relay allowing you to control up to 220V and 20A (see Figure 12.5). However, it lacks some of the features that make the PowerSwitch Tail and Simple Relay Shield shine:

- The Beefcake doesn't have the great shield configuration that makes the Simple Relay Shield convenient because it connects directly to headers on the Arduino.

FIGURE 12.5 SparkFun's Relay Board can control 20 amps and 220 volts.

- Similarly, it lacks the ease of use and safety protection afforded by the PowerSwitch Tail. The Beefcake Relay Control Board is only recommended for those experienced in using high voltage safely.

SAFETY: ELECTRICITY

As everyone knows—or ought to know—electricity can hurt or even kill you, damage electrical equipment, and can cause fires that destroy property and harm people. It is imperative, therefore, that you handle high-voltage electricity with extreme care, or better yet, don't handle it at all! The symbol shown in Figure 12.6 warns you of the potential electrical hazard. Just don't assume that a symbol such as this will always be present when the risk for electrical shock is present.

FIGURE 12.6 When you see this warning symbol, you'll know there's an electrical hazard nearby.

Keep the following safety tips in mind at all times:

- Avoid contact with bare wires and exposed terminals, including those on ostensibly HV-rated circuit boards. Anything more than 50V should have an enclosure or other insulation to prevent accidental contact.
- Treat all electrical devices as if they were live or energized. Test using a voltmeter if you're not sure. Also, be aware that some components, such as capacitors, retain a charge for a long time after the part has been connected to a power source.
- Disconnect the power source before working on any piece of equipment.
- Avoid conductive tools, jewelry, and other items that could transmit electricity to your body.
- Don't use electrical equipment—including power cords—that have been damaged or improperly modified.
- Don't use electrical equipment that is wet, whether submerged or simply dripping. Unplug the equipment and let it dry out before you work on it. Even heavy condensation can transmit a lethal shock!
- Do not attempt to touch, repair, or open a high-voltage project unless you really, REALLY know what you're doing!

Mini Project: Making a Fan Controller

For the mini project, you'll build a fan controller that starts a fan when the temperature reaches a certain level (see Figure 12.7). The controller consists of an Arduino Uno, a temperature sensor, and a PowerSwitch Tail, with the latter connecting an ordinary desk fan to house current.

FIGURE 12.7 Turn on a fan when the temperature reaches a certain point.

PARTS LIST

You need just a couple of things for this project:

- Arduino Uno and wall wart
- PowerSwitch Tail II (Adafruit P/N 268)
- LM355AZ temperature sensor (Jameco P/N 120820)
- Breadboard
- Jumpers
- A fan that operates on 110V

Instructions

This is a quick-and-dirty build, with only five wires and a single electronic component, shown in Figure 12.8.

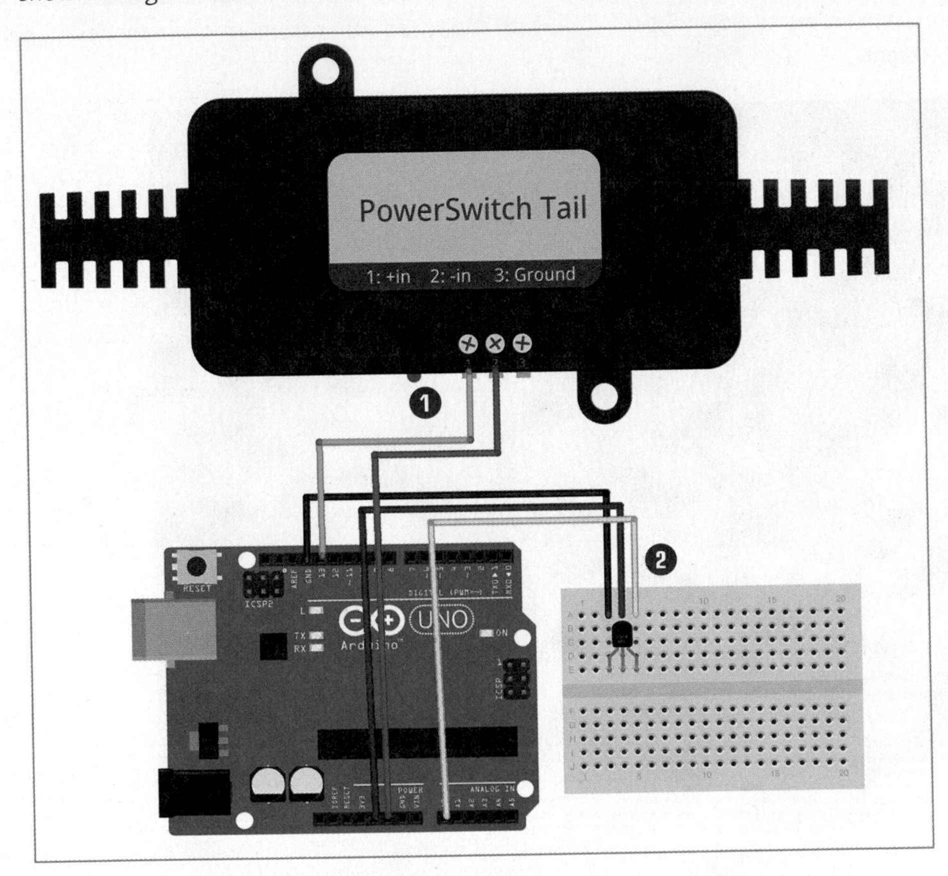

FIGURE 12.8 The fan controller is a quick and easy build.

1. Connect the first and second terminals of the PowerSwitch Tail to pin 13 and a GND on the Arduino. In Figure 12.8, these are shown as being green and orange wires, respectively.
2. Looking at the flat face of the temperature sensor, connect the left lead to GND (black wire), the center lead to 5V (red wire), and the right-hand lead (yellow) to A0.

Finally, plug the male end of the PowerSwitch Tail into an electrical outlet and connect the female end to the fan's plug.

Fan Controller Code

Upload the following sketch to the Arduino. As always, if you can't remember how to do it, I explain how to upload code in Chapter 5, "Programming Arduino."

> **NOTE**
>
> **Code Available for Download**
>
> You don't have to enter all of this code by hand. Simply go to https://github.com/n1/Arduino-For-Beginners to download the free code.

```
int sensorPin = A0;     // connect the data pin of the sensor here
int fanPin = 13;        // connect the PowerSwitch Tail here
int sensorValue = 0;   // variable to store the value coming from the sensor

void setup() {

    pinMode(fanPin, OUTPUT);
    Serial.begin(9600);
    Serial.println("starting!");
}

void loop() {
  sensorValue = analogRead(sensorPin);

   Serial.print(sensorValue);

// Change the sensorValue number here depending at what temperature
// you want the fan to start.
   if (sensorValue >= 753)
{
  digitalWrite(fanPin, HIGH);
    delay(10000); // how long the fan stays on in milliseconds
}
   else
 {
  digitalWrite(fanPin, LOW);
  delay(10000); //  how long before the sensor checks again in MS
 }
}
```

Project: Making a Lava Lamp Buddy

Everyone loves lava lamps, those friendly glowing cones of bubbling liquid. They're actually very simple: a light bulb concealed in the base both lights up and heats a jar of wax and liquid. When the wax reaches a certain temperature, it starts bubbling and moving around.

One downside of lava lamps is that they take a while to heat up. When it occurs to you that you would like to have the lamp on, and flick the switch, the lamp still needs a good hour until it gets interesting to look at. Wouldn't it be awesome if you could set a schedule so your lamp turns on automatically an hour before you get home from work? Additionally, many people don't realize that manufacturers recommend keeping your lamp on no more than 10 hours at a time, so the lamp should shut off automatically as well. Finally, we're all lazy, and being able to use an ordinary TV remote control to turn on and off the lamp would be perfect.

All you need is an Arduino-controlled Lava Lamp Buddy (see Figure 12.9) to control the lamp's schedule and interface with a remote control. As luck would have it, that is precisely this chapter's project!

FIGURE 12.9 The Lava Lamp Buddy controls your lava lamp so you don't have to!

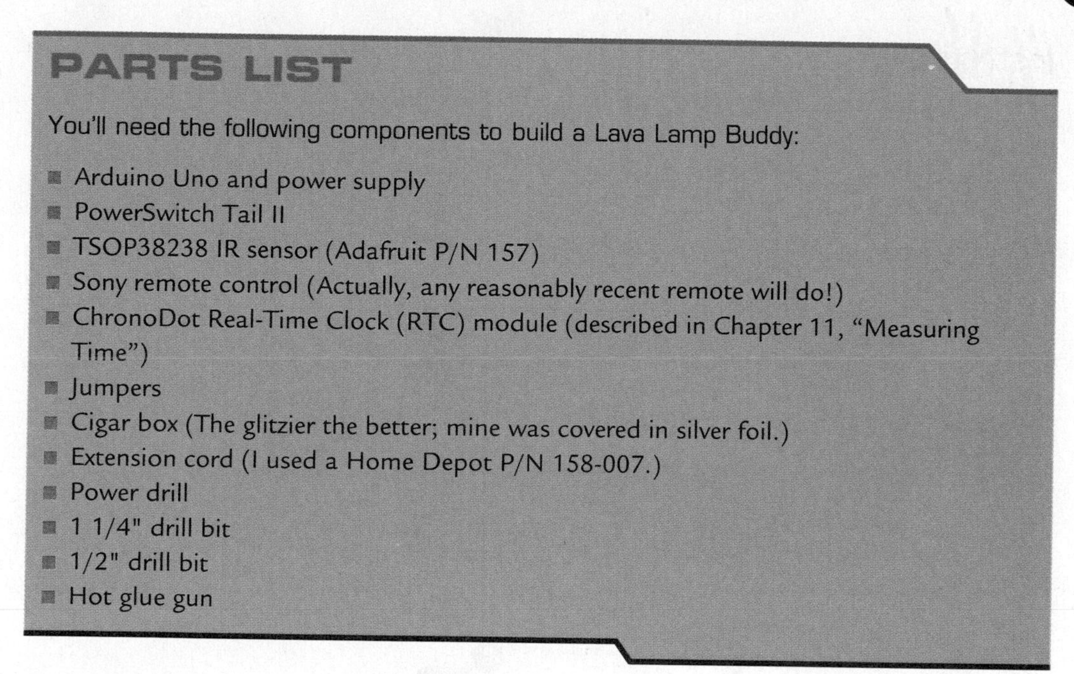

PARTS LIST

You'll need the following components to build a Lava Lamp Buddy:

- Arduino Uno and power supply
- PowerSwitch Tail II
- TSOP38238 IR sensor (Adafruit P/N 157)
- Sony remote control (Actually, any reasonably recent remote will do!)
- ChronoDot Real-Time Clock (RTC) module (described in Chapter 11, "Measuring Time")
- Jumpers
- Cigar box (The glitzier the better; mine was covered in silver foil.)
- Extension cord (I used a Home Depot P/N 158-007.)
- Power drill
- 1 1/4" drill bit
- 1/2" drill bit
- Hot glue gun

Decoding Infrared

Infrared sensors (see Figure 12.10) are obviously designed to notice infrared light, but they are very selective. Only infrared (IR) light pulsing at 38 Khz (that's 38,000 off-and-on cycles per second) is sensed, and the sensor toggles its voltage output accordingly. If it detects a 38 Khz carrier, it outputs 0V; otherwise, it outputs 5V.

FIGURE 12.10 An infrared sensor like this one listens for signals of 38 Khz but ignores all other infrared light. Credit: Adafruit Industries

The 38Khz number brings up the opposing problem: How do you send such a signal? You can use a number of electronic tricks involving pulsing an infrared LED, but many tinkerers use ordinary household remote controls. Each button's IR code can be scanned in and the programmer can tell the Arduino to perform a different action for each code. In other words, you could put an IR sensor on a robot and control the bot with the same remote control you use with your TV.

Instructions

This is a slightly more complicated rig than the mini project earlier in this chapter, which also features a PowerSwitch Tail. Not only are you adding an IR sensor, but there's a real-time clock module, as well as the expected Arduino. Here's how to wire up the Lava Lamp Buddy, following along with Figure 12.11.

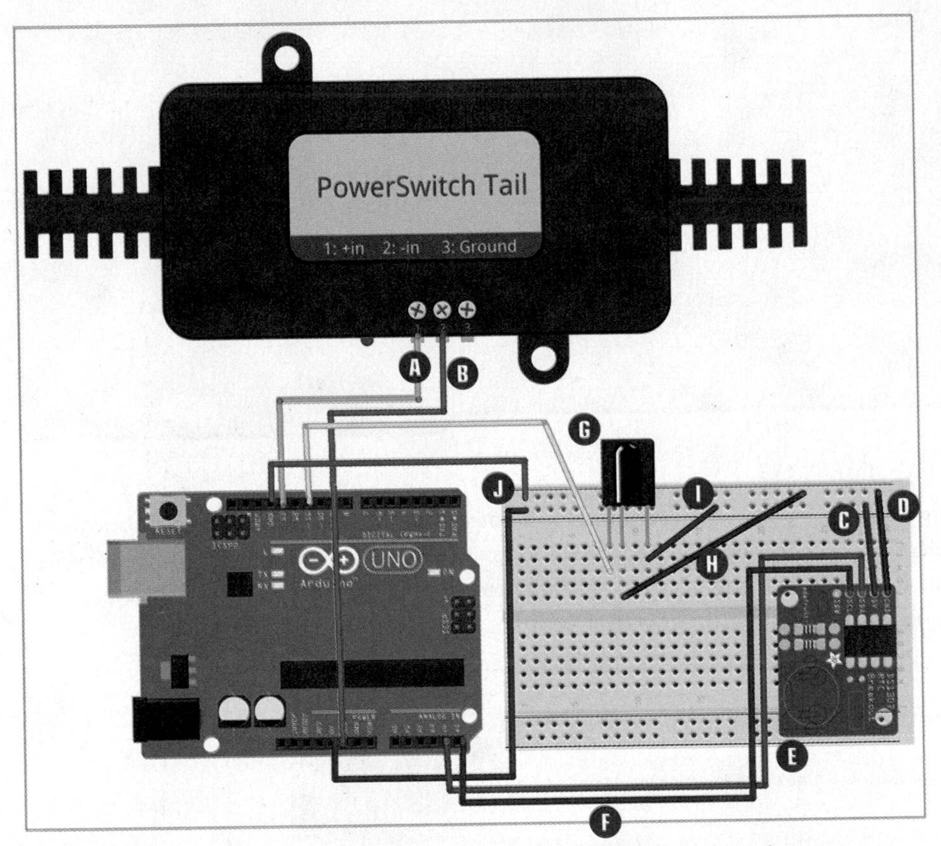

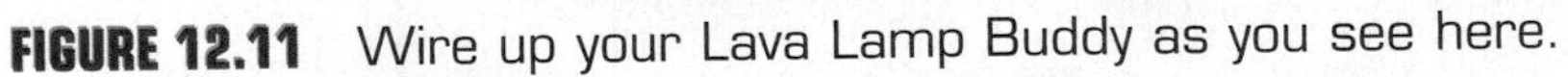
FIGURE 12.11 Wire up your Lava Lamp Buddy as you see here.

1. Wire up the Arduino, PowerSwitch Tail, RTC module, and IR sensor, as you see in Figure 12.11.

 Connect terminal 1 of the PowerSwitch Tail to pin 13 of the Arduino. This is the green wire shown in Figure 12.11.

 Connect terminal 2 to a GND pin on the Arduino. This is the orange wire shown in Figure 12.11.

 Plug in the RTC module to the breadboard and connect the 5V pin (red wire) to the breadboard's power bus.

D Connect the GND (black wire) to the breadboard's ground bus.

E Connect the SDA (purple wire) to A4 on the Arduino.

F Connect the SCL (brown wire) to A5 on the Arduino.

G Solder jumpers (see Figure 12.12) to the infrared sensor so it can be attached to the outside of the cigar box.

H The infrared sensor's left lead (yellow wire), looking at the bulb on the component's face, connects to pin 11 on the Arduino.

I The infrared sensor's center lead (black wire) goes to the breadboard's ground bus.

J The infrared sensor's right-hand lead (red wire) plugs into the breadboard's power bus.

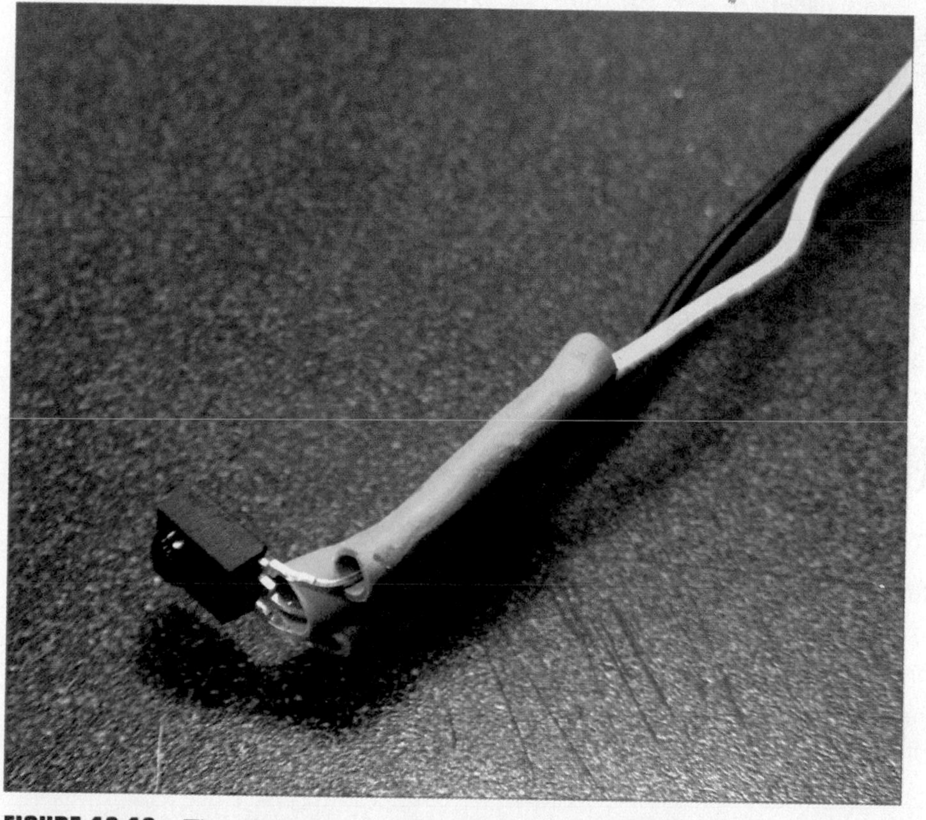

FIGURE 12.12 The IR sensor with jumpers soldered on.

2. Drill a hole in the back of your cigar box with the 1 1/4" bit, and drill a hole in the front of the box with the 1/2" bit.
3. Hot glue the sensor to the front of the box with the wires passing through the hole, as shown in Figure 12.13.

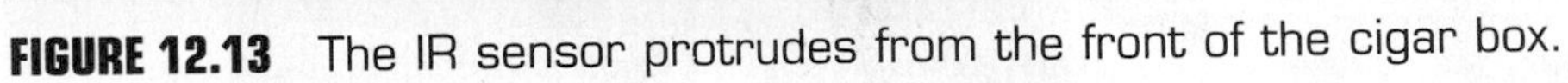

FIGURE 12.13 The IR sensor protrudes from the front of the cigar box.

4. Hot glue the breadboard and Arduino in place inside the cigar box.
5. Plug in the Arduino's power supply and the PowerSwitch Tail into the extension cord. The cord I specified has three plugs in the end; you can just use a splitter if you want.
6. Pass the lava lamp's power plug through the hole in the back of the cigar box so it can plug into the PowerSwitch Tail's female end. It should look more or less like Figure 12.14.

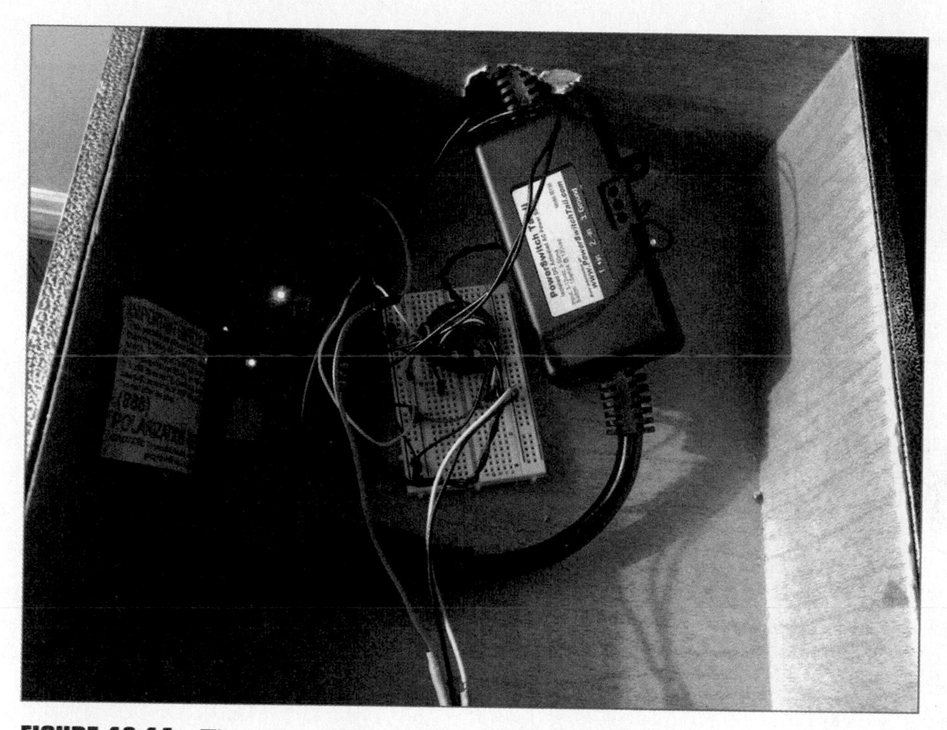

FIGURE 12.14 The guts of the Lava Lamp Buddy.

Lava Lamp Buddy Code

This code is rather complicated because it's doing two things:

- Interpreting IR signals from the remote control
- Pulling in data from the RTC

Assisting in this work are three libraries, which you'll have to download before the sketch will upload.

- The wire.h library comes pre-installed with Arduino, so you don't have to worry about this one.
- The RTC library is available from Adafruit's github repository: https://github.com/adafruit/RTClib.
- Finally, you can find the IRremote.h library in Ken Shirriff's github: https://github.com/shirriff/Arduino-IRremote/blob/master/IRremote.h.

NOTE

Code Available for Download

You don't have to enter all of this code by hand. Simply go to https://github.com/n1/Arduino-For-Beginners to download the free code.

```
#include <Wire.h>
#include "RTClib.h"
#include <IRremote.h>

int RECV_PIN = 11;
int RELAY_PIN = 13;
int startTime = 1600; // lamp turns on at this time -- 1600 is 4pm.
int stopTime = 2200; // lamp turns off at this time -- preset for 2200 or 10pm.

RTC_Millis RTC;

IRrecv irrecv(RECV_PIN);
decode_results results;

void dump(decode_results *results) {
  int count = results->rawlen;
  if (results->decode_type == UNKNOWN) {
    Serial.println("Could not decode message");
  }



  for (int i = 0; i < count; i++) {
    if ((i % 2) == 1) {
      Serial.print(results->rawbuf[i]*USECPERTICK, DEC);
    }
    else {
      Serial.print(-(int)results->rawbuf[i]*USECPERTICK, DEC);
    }
    Serial.print(" ");
  }
  Serial.println("");
}

void setup()
```

```
{
        Serial.begin(57600);
      // following line sets the RTC to the date & time this sketch was compiled
      RTC.begin(DateTime(__DATE__, __TIME__));


    pinMode(RELAY_PIN, OUTPUT);
    pinMode(RECV_PIN, INPUT);
      Serial.begin(9600);
    irrecv.enableIRIn(); // Start the receiver
}

int on = 0;
unsigned long last = millis();

void loop() {


        DateTime now = RTC.now();

        int decimalTime = now.hour() * 100 + now.minute();
        Serial.print(decimalTime);
delay (1000);

      Serial.println();


      if (decimalTime == startTime) {
            digitalWrite(RELAY_PIN, HIGH);
      }

      if (decimalTime == stopTime) {
            digitalWrite(RELAY_PIN, LOW);
      }


   if (irrecv.decode(&results)) {
     // If it's been at least 1/4 second since the last
     // IR received, toggle the relay
     if (millis() - last > 250) {
       on = !on;
       digitalWrite(RELAY_PIN, on ? HIGH : LOW);
```

```
      dump(&results);


    }
    last = millis();
    irrecv.resume(); // Receive the next value

  }
}
```

The Next Chapter

Chapter 13, "Controlling Motors," introduces you to various techniques for controlling motors. Whether they're steppers, servos, or regular old DC motors, you'll learn how to control them with the help of your Arduino. You'll then take what you've learned and build a fabulous bubble-making machine.

13

Controlling Motors

In this chapter, you'll add to your motor knowledge by exploring motor control boards, which enable you to control and power all sorts of motors. You'll then work on this chapter's project, a BubbleBot that spreads joy and soap bubbles throughout the neighborhood (see Figure 13.1).

FIGURE 13.1 Need more bubbles in your life? Okay, silly question. Of course you do.

How to Control Motors

We've already covered the basic motor types—steppers, servos, and DC motors. Now let's talk about how you control them using an Arduino! The secret is that you have to use a motor control chip such as the L293D. It manages the flow of data between the motor and the Arduino, enabling you to control more motors than you would ordinarily be able to.

Even better, motor control chips are the brains of convenient motor control boards that include extra features such as supplying power to the individual motors. Let's examine three cool examples of motor control boards.

Adafruit Motor Shield

The Motor Shield (Adafruit P/N 81; see Figure 13.2) is kind of a perfect weapon for controlling motors. One of the biggest limitations when running motors from an Arduino is running out of pins and power, because motors use a lot of both. For instance, a servo motor uses three wires, each of which would ordinarily need its own pin. Furthermore, the Arduino's 5V pin can barely handle one servo much less a number of them. The motor shield manages power and data so only the bare minimum of resources are needed. It can run two servos and two steppers, or up to four DC motors in place of the steppers. All of this with the convenient shield form factor.

FIGURE 13.2 The Adafruit Motor Shield—the circuit board in the middle—can control DC motors, steppers, and servos. Credit: Adafruit Industries

Shmalz Haus EasyDriver

A more elegant solution than a full-fledged shield, the EasyDriver (SparkFun P/N ROB-10267, pictured in Figure 13.3) stepper controller is a single board with inputs for data and power, with an on-board voltage regulator controlling how much juice your stepper gets. It only costs around $15 and is far smaller than an Arduino shield.

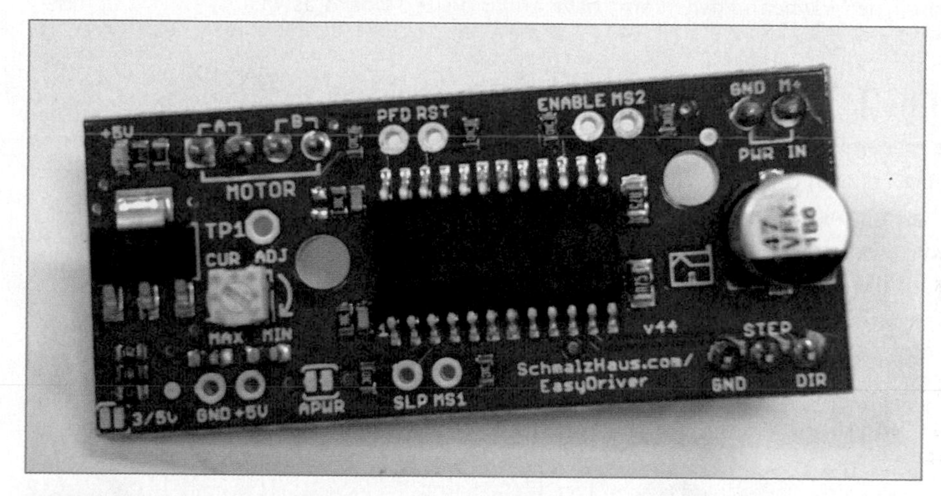

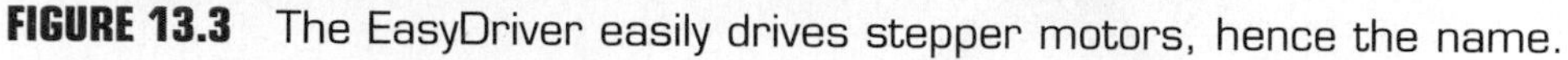

FIGURE 13.3 The EasyDriver easily drives stepper motors, hence the name.

Bricktronics MegaShield

Featuring three L293D chips and able to control six motors and take input from four sensors, the Bricktronics MegaShield (see Figure 13.4) allows you to control LEGO Mindstorms motors, even accommodating LEGO's proprietary cables.

FIGURE 13.4 The Bricktronics MegaShield controls up to six motors and takes input from as many as four sensors.

It's a fairly robust board, exceeding the specs for LEGO's own Mindstorms microcontroller while offering all the programmability of the Arduino platform.

Note the term "MegaShield." It's a reference to an Arduino Mega, a really big Arduino with a lot more computing power than an Uno. So basically, this board is a shield for a Mega and wouldn't work with an Uno, the Arduino we use in this book. Never fear, designers Wayne and Layne (wayneandlayne.com) have an Uno-sized board as well.

Powering Your Motor Using a TIP-120

Part of the reason why motor control boards exist is because the Arduino has a hard time powering motors with only its on-board power supply, which consists of 3.3V and 5V for the two relevant pins. That might be enough for one motor, but for a robot with several motors, your average Arduino won't be able to keep up. One solution might be a motor control board like the ones mentioned at the beginning of this chapter. A simpler and cheaper alternative is to use a Darlington transistor, like the TIP-120 (see Figure 13.5; Adafruit P/N 976).

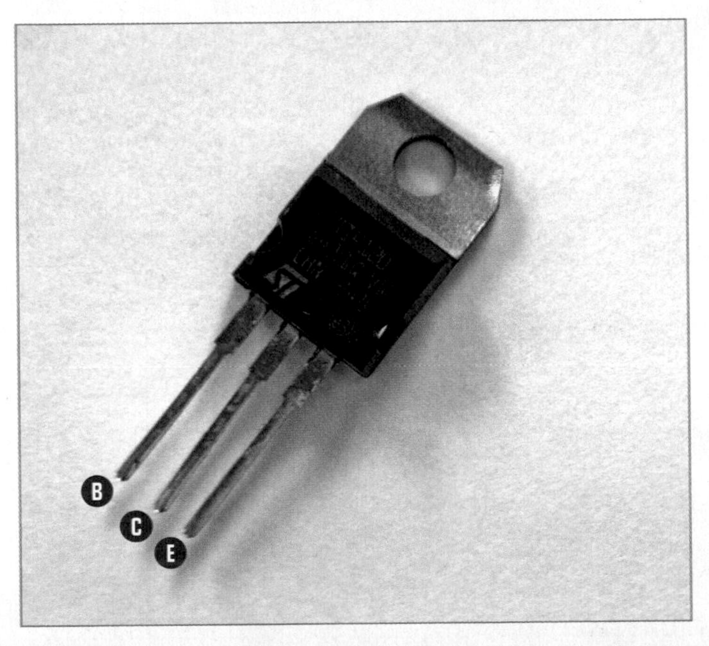

FIGURE 13.5 A Darlington transistor controls electricity so your Arduino doesn't have to.

A Darlington transistor is basically a solid-state, electrically actuated switch, allowing you to control larger amounts of electricity with a tiny bit of current.

Here's how it works. The transistor has three terminals protruding from it, and these are called the base, collector, and emitter, often abbreviated B, C, and E. You can see these marked on Figure 13.5.

- **Base**—This pin triggers the circuit when it gets pinged by the Arduino.
- **Collector**—You hook up your power supply to the middle pin.
- **Emitter**—Power from the collector is released by the emitter when commanded by the base.

Figure 13.6 shows a simple example of how you would wire up an LED to turn on when pinged by the Arduino.

1. The power supply is wired up to the positive lead of the LED with a 220-ohm resistor in between.

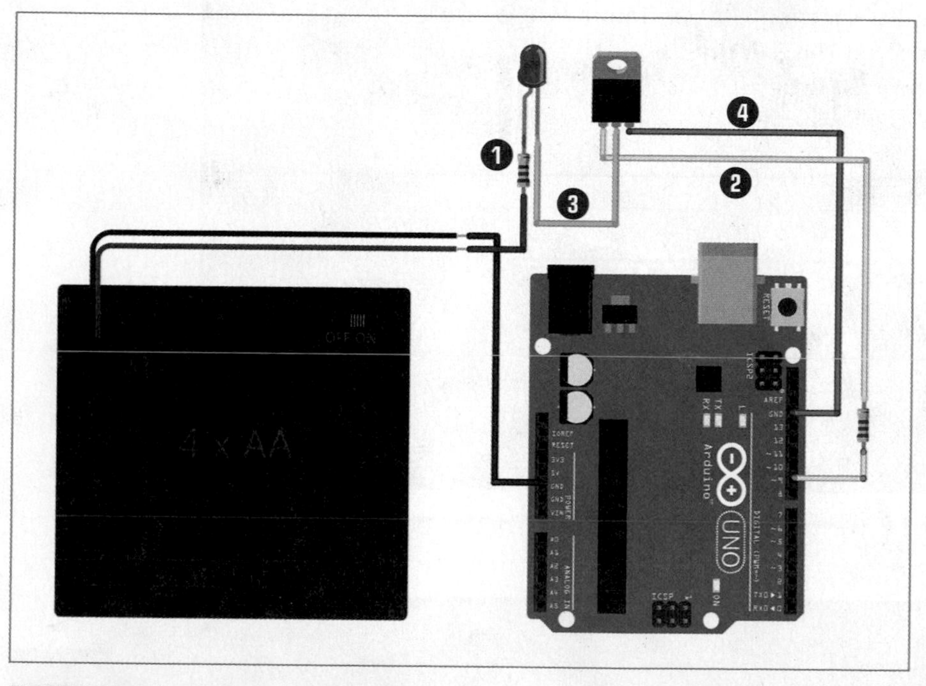

FIGURE 13.6 Wiring up a TIP-120.

 The base is connected to a digital pin of the Arduino; this pin will trigger the circuit.

3. Where it gets weird is that the negative lead of the LED is connected to the collector of the transistor.

4. The emitter, which is supposed to release the voltage when triggered, goes to ground. What gives? Think of it this way: The base of the transistor is the trigger, and the collector and emitter are the circuit. When the base gets pinged, the entire loop beginning at the battery pack, passing through the resistor, LED, transistor, and then to ground, immediately becomes a circuit.

Alt.Project: Stepper Turner

In this project, you'll set up a stepper motor to turn as directed by a potentiometer (see Figure 13.7). As you learned in Chapter 1, "Arduino Cram Session," a potentiometer (or pot, as it's called) is an analog device that delivers a variable amount of resistance depending on how far the knob is turned. You can take a reading from the pot and turn it into a value (degrees) that can be used to direct how far the motor turns.

FIGURE 13.7 Turn a stepper motor as directed by a potentiometer.

PARTS LIST

You'll need the following parts to build the project:

- Arduino Uno
- 12V power supply (You can use an 8 × AA battery pack if you don't have another type of power supply.)
- Schmalz Haus EasyDriver (described previously in this chapter)
- Stepper motor (Adafruit P/N 858)
- Potentiometer (You can pretty much use any one; try Adafruit P/N 562.)
- Jumpers (the usual!)
- Breadboard

Instructions

This project consists of a surprising number of connections, but just follow along with Figure 13.8.

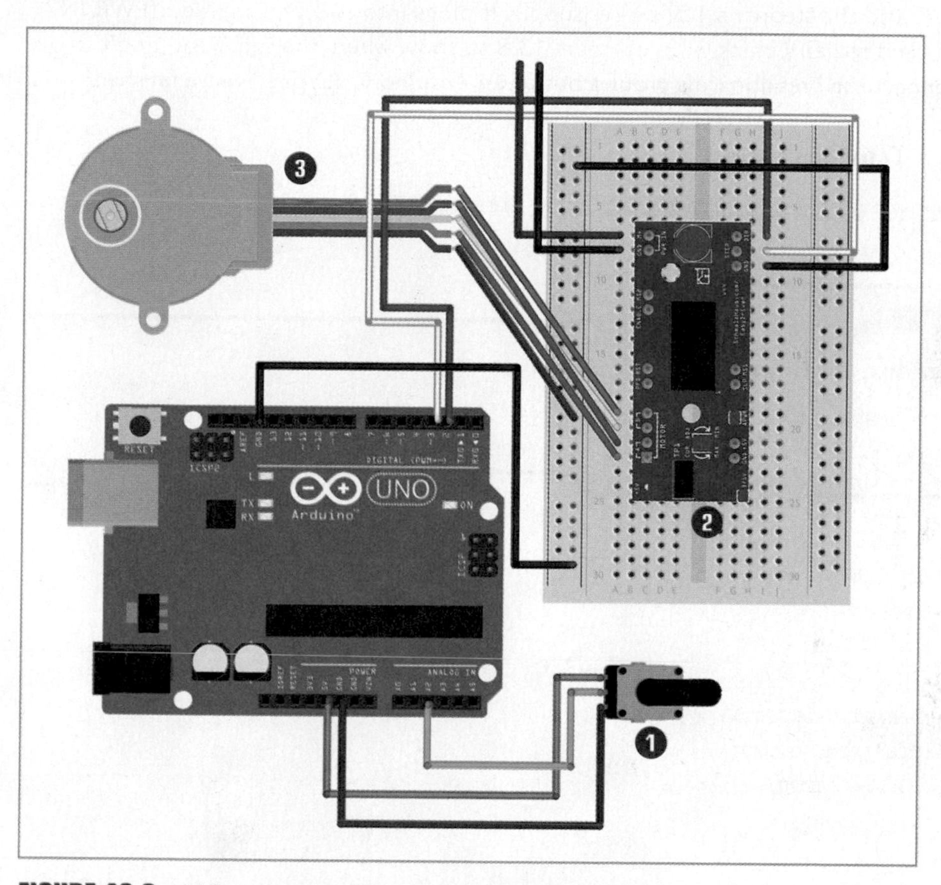

FIGURE 13.8 Wire up the project as you see here.

1. Connect the potentiometer. The left terminal, with the terminals pointed toward you, connects to 5V on the Arduino (the gray wire), the middle one to A2 (green), and the right terminal (brown) to GND.
2. Plug in your EasyDriver to a breadboard. Connect it to the Arduino with two jumpers, marked as white and purple in Figure 13.8. The white wire connects the pin marked “STEP” on the PCB to pin 3 on the Arduino. The purple wire connects from the pin marked “DIR” on the PCB to pin 2 on the Arduino. The pin marked GND on the EasyDriver plugs into the ground bus.
3. The stepper motor has five wires: red, orange, yellow, pink, and blue. The red wire is ground and can plug into the breadboard’s ground bus. The other four wires plug into

the four pins labeled "MOTOR" on the EasyDriver PCB. However, it's a little tricky. You can't just plug them in in the same order the yellow and pink wires need to be swapped, as marked in Figure 13.8.

4. Finally, add the stepper's 12V power supply. It plugs into two pins marked "PWR IN," and I used red and black wires in Figure 13.8 to show where they go. Also, don't forget to connect the breadboard's ground bus to an Arduino GND pin. You're finished!

Stepper Turner Code

Use the following code to program the Stepper Turner.

> **NOTE**
>
> **Code Available for Download**
>
> You don't have to enter all of this code by hand. Simply go to https://github.com/n1/Arduino-For-Beginners to download the free code.

```
#define DIR_PIN 2
#define STEP_PIN 3
#define potPin A2

void setup() {
  pinMode(DIR_PIN, OUTPUT);
  pinMode(STEP_PIN, OUTPUT);
  pinMode(potPin, INPUT);
  Serial.begin(9600);
}

void loop(){

int potReading = analogRead(potPin);

Serial.println(potReading);

  rotateDeg(potReading, 1);
  delay(1000);
}
```

```
void rotate(int steps, float speed) {

  int dir = (steps > 0)? HIGH:LOW;
  steps = abs(steps);

  digitalWrite(DIR_PIN,dir);

  float usDelay = (1/speed) * 70;

  for(int i=0; i < steps; i++){
    digitalWrite(STEP_PIN, HIGH);
    delayMicroseconds(usDelay);

    digitalWrite(STEP_PIN, LOW);
    delayMicroseconds(usDelay);
  }
}

void rotateDeg(float deg, float speed){

  int dir = (deg > 0)? HIGH:LOW;
  digitalWrite(DIR_PIN,dir);

  int steps = abs(deg)*(1/0.225);
  float usDelay = (1/speed) * 70;

  for(int i=0; i < steps; i++){
    digitalWrite(STEP_PIN, HIGH);
    delayMicroseconds(usDelay);

    digitalWrite(STEP_PIN, LOW);
    delayMicroseconds(usDelay);
  }
}
```

Project: BubbleBot

For the final project of this book, you're going to kick summer (or whatever season it might be as you're reading this) into high gear with this excellent BubbleBot, shown in Figure 13.9. It's a simple robot with a wooden framework, and it dips a bubble wand into a tray of bubble solution, then raises it up and blows on it. It keeps blowing bubbles until it runs out of soap or you pull the plug!

FIGURE 13.9 Beauty shot—the BubbleBot.

PARTS LIST

Gather the following supplies to build your BubbleBot:

- Arduino Uno
- Adafruit motor shield (P/N 81)
- Servo (I used a HiTec HS-322HD servo, Jameco P/N 33322.)
- Servo horns (The ones that came with the servo are fine.)
- Switch (Jameco P/N 76523)
- Mini breadboard (SparkFun P/N 11658)
- 12V battery pack (Adafruit P/N 449)

- Computer fan (I used a Comair P/N FE24B3 fan.)
- 1/4" MDF for chassis; an 18" × 24" sheet should suffice.
- A bubble wand (I used a wand from a 25-piece Miracle Bubbles set.)
- 1/2" diameter wooden dowel, about 10" in length
- 2 1/4" threaded rods, each about 8" long (You can buy threaded rods at any hardware store.)
- A couple of #4-40 × 3/8" wood screws
- 8 1/4" nuts with locking washers
- 8 #4-40 × 1" machine screws with washers and nuts
- Jumpers (the same sort you've used throughout the book)
- Wood glue
- Hot glue and hot glue gun

Instructions

After you have gathered the parts together, it's time to begin building! Follow along with these steps to create your BubbleBot:

1. Laser-cut the enclosure out of quarter-inch MDF, shown in Figure 13.10. You can download the design from https://github.com/n1/Arduino-For-Beginners. Alternatively, simply build a wooden chassis as you normally would and drill the holes for the hardware.

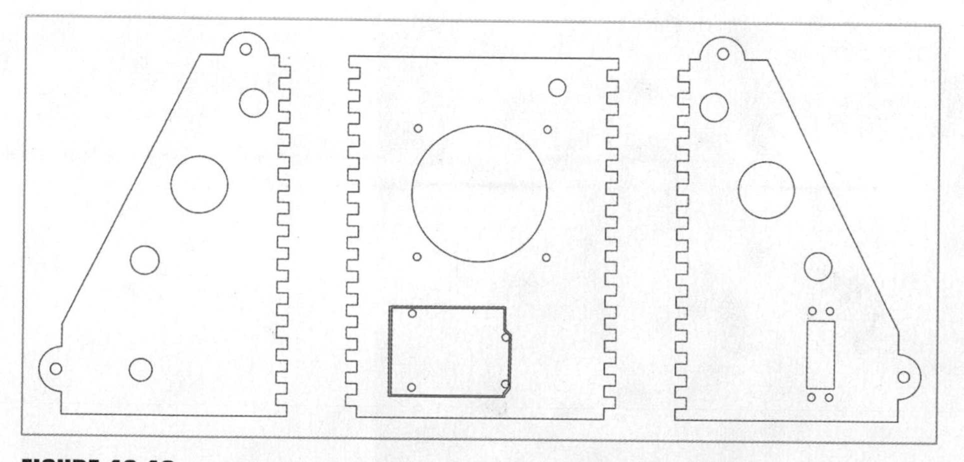

FIGURE 13.10 Laser-cut this design to make the BubbleBot's chassis.

2. Mesh the teeth on the sides and back of the chassis and glue them in place with wood glue.

TIP

My Prototype Looks a Little Different

When I built my own BubbleBot, my prototype used screws instead of teeth to hold the sides to the back. Having built the robot, I redesigned the chassis somewhat to take advantage of the techniques I learned the first time around. So basically, there may be a couple of cosmetic differences between the two versions but mostly they are the same.

3. Insert the threaded rods with locking washers and nuts, as shown in Figure 13.11. Tighten the hardware so the sides of the chassis don't move.

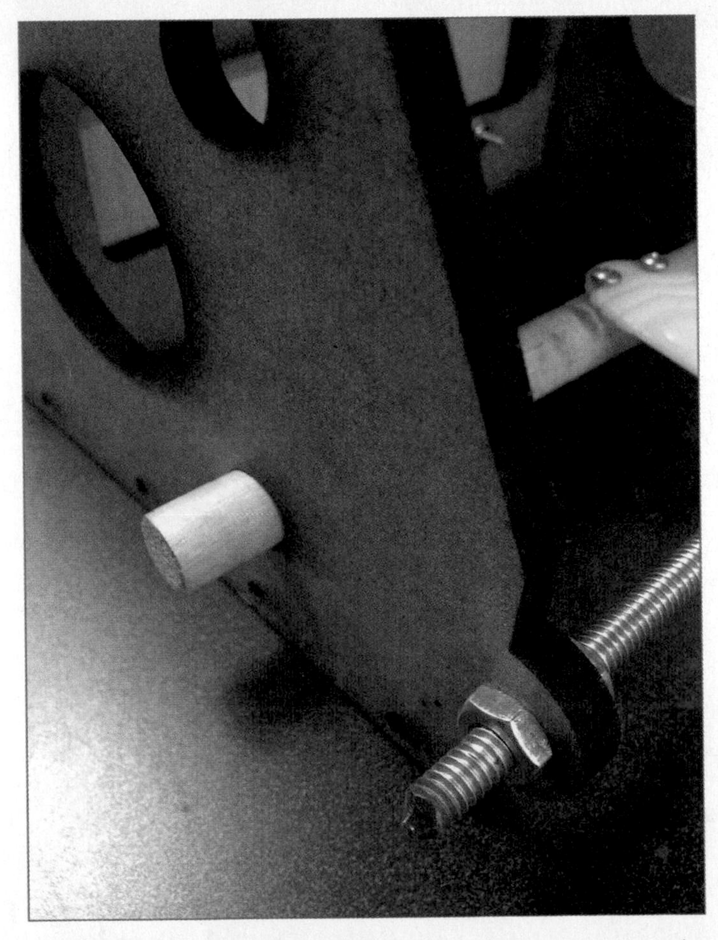

FIGURE 13.11 The threaded rods help secure the chassis.

4. Use the 3/8" #4 screws to connect the servo horn to one end of the dowel, as shown in Figure 13.12.

FIGURE 13.12 Secure the servo horn to the end of the dowel.

5. Insert the dowel into the hole opposite the servo's horn. It doesn't need to be secured because the opposite end is connected to the servo. Figure 13.13 shows how it should look. If it sticks out too much, feel free to trim it down.

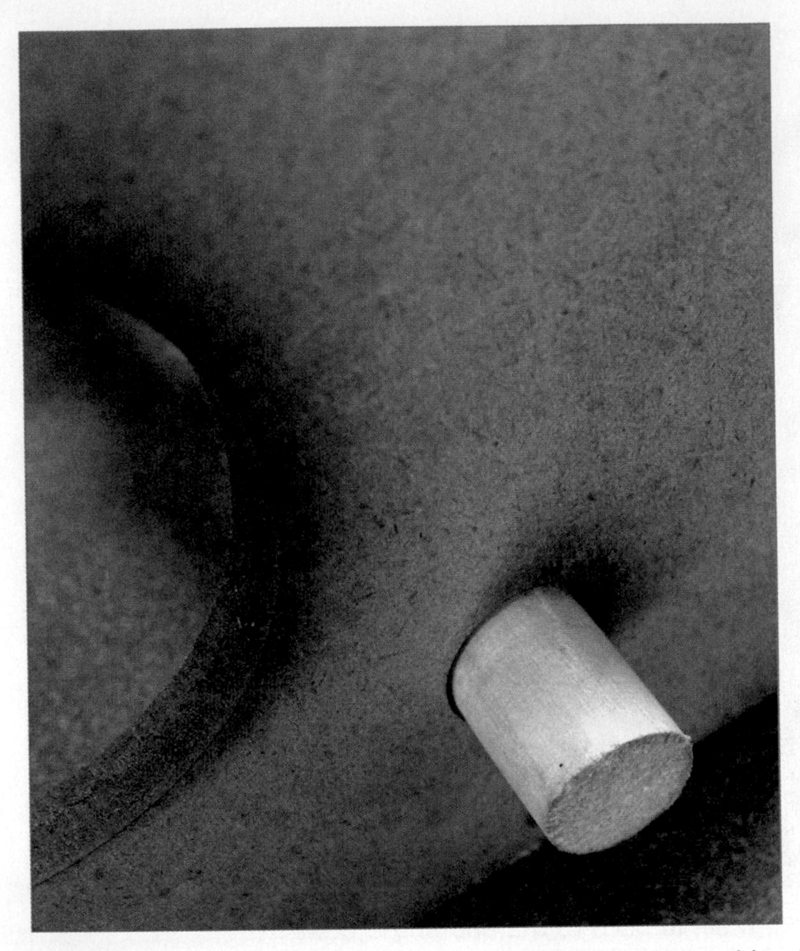

FIGURE 13.13 The dowel protrudes from the opposite side of the chassis.

6. Drill pilot holes and attach the bubble wand to the dowel with the 3/8" #4 wood screws, as shown in Figure 13.14.

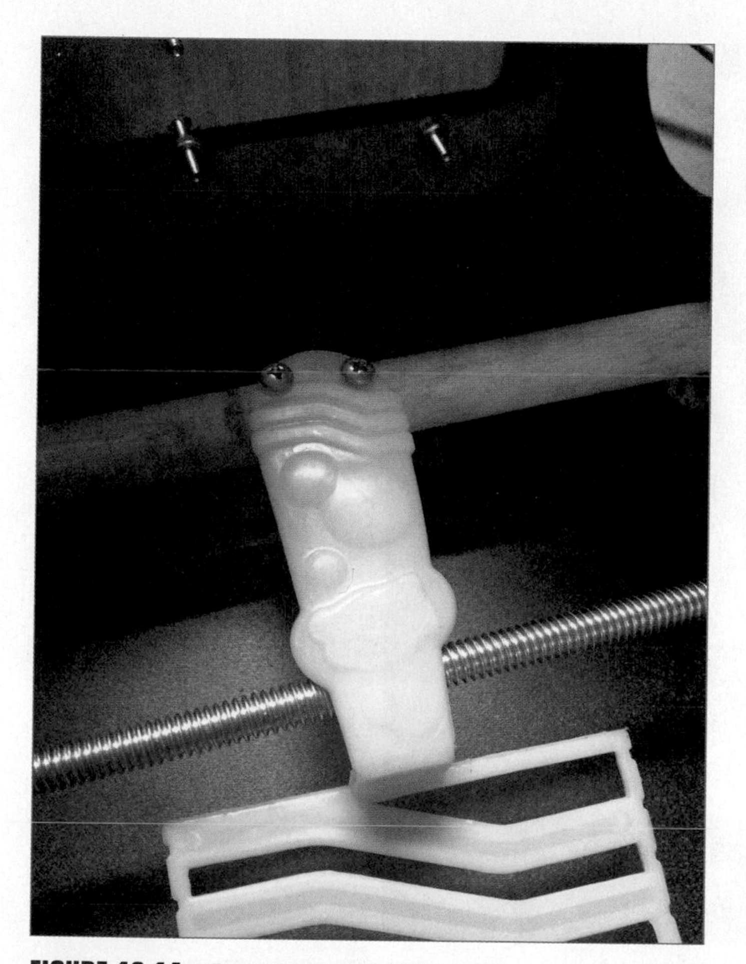

FIGURE 13.14 Screw on the bubble wand to the dowel.

7. Thread the switch through the small hole in the upper portion of the back panel. Secure it with the hex nuts that came with the switch, and thread the wires through the fan hole, as shown in Figure 13.15.

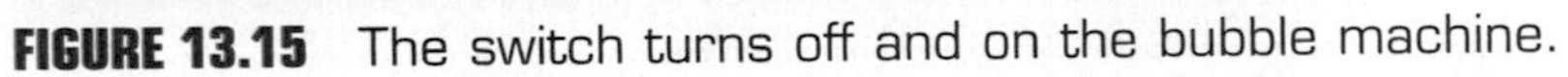

FIGURE 13.15 The switch turns off and on the bubble machine.

8. Attach the computer fan (see Figure 13.16) to the back of the chassis using the #4 × 1" screws, washers, and nuts.

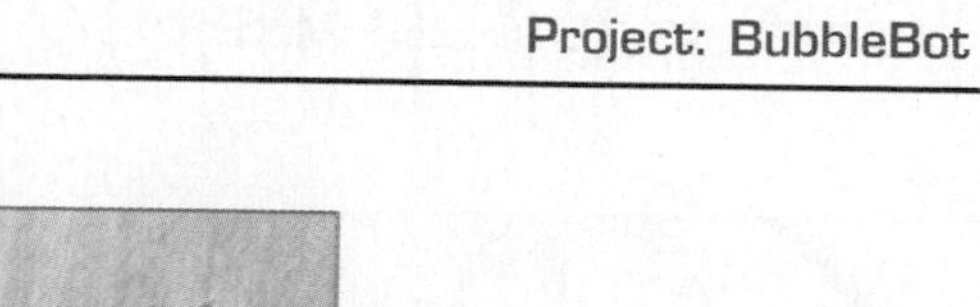

FIGURE 13.16 Attach the computer fan to the back panel of the enclosure.

Attach the Arduino to the back panel using the 1" #4 screws, and then mount the Adafruit motor shield to the Arduino by inserting the pins of the shield into the Arduino's headers. Wire up the shield, following along with Figure 13.17.

FIGURE 13.17 The motor shield controls your BubbleBot.

1. Solder wires to the switch if you haven't already, then connect the black wire to the 3V pin on the motor shield, and the red wire to pin 2. You might want to add just a dab of solder to keep the wires from falling out of the shield.
2. Plug in the servo's wires to the pins marked "SERVO2." The red wire should plug into +, the black to –, and the yellow to "S". Note that servos combines their wires into triple plugs, so this is totally foolproof.
3. Connect the fan's wires to the blue terminal block marked "M2." The red and black wires can go either way.
4.  Hot glue the 12V battery pack to the back or side of the chassis. Connect the battery pack to the blue terminal block marked "+M" and "GND." This will power the motors.

9. Hot glue the 9V battery pack (not shown in Figure 13.17) to the chassis and connect the plug to the Arduino's power plug. This will power the Arduino.

BubbleBot Code

Upload the following code to your BubbleBot's Arduino. Note that you'll need to download Adafruit's AFMotor.h library from its code repository: https://github.com/adafruit/Adafruit-Motor-Shield-library.

NOTE

Code Available for Download

You don't have to enter all of this code by hand. Simply go to https://github.com/n1/Arduino-For-Beginners to download the free code.

```
// This code is based on Adafruit's Motor Shield example code.

#include <AFMotor.h>
#include <Servo.h>


AF_DCMotor motor(2);
Servo servo1;
int toggle = 2;
int toggleStatus = 0;



void setup() {
  Serial.begin(9600);

      pinMode(toggle, INPUT);

  // turn on servo
  servo1.attach(9);

  // turn on motor #2
  motor.setSpeed(200);
  motor.run(RELEASE);
}

int i;

void loop() {

  int toggleStatus = digitalRead(toggle);
  if (toggleStatus == 1) {

     Serial.println(toggleStatus);
```

```
motor.run(FORWARD);
for (i=255; i!=0; i--) {
  servo1.write(i-90
  );
  delay(30);
}

delay(5000);
Serial.println("waiting1");

motor.run(FORWARD);
for (i=0; i<2000; i++) {
  motor.setSpeed(i);
  delay(5);
}

delay(5000);

}
}
```

Glossary

3D printer—A machine able to extrude and deposit layers of plastic in order to form a three-dimensional object.

analog—Data sent in a continuous wave of varying voltage, as opposed to digital, which sends data with a series of on-and-off signals.

array—In programming terminology, an array is a list of values stored for future use.

band saw—A power saw, used for woodworking and metalworking, consisting of a loop-shaped saw blade.

Barbot—A robot designed to make and serve cocktails.

barometric sensor—A sensor that detects changes in air pressure, much the way a barometer does.

Baud rate—The speed in which data is transmitted; baud value equates to the number of characters sent per second. So, 9600 baud equals 9,600 characters transmitted every second.

bit—One piece of data, usually assumed to be a 0 or 1.

Bluetooth—A low-power, wireless data protocol used by computer mice, wireless earphones, and other commercial applications.

board—A shorter way of saying a PCB, or printed circuit board.

breadboard—A hole-punched plastic board with concealed conductors, allowing you to wire up circuits easily and without solder.

breakout board—A small PCB used for controlling a single component. For instance, you could create a breakout board for managing an L293D motor control chip.

caliper—A device for accurately measuring short distances.

capacitor—An electronic component that stores small amounts of electricity in an electrostatic field.

circuit bending—A technique for retrofitting commercial electronic toys and devices to change their behaviors.

compile—To convert one computer language to another, typically used to turn people-readable code to machine-readable code.

Computer Numerically Controlled (CNC) tools—Rail-mounted power tools that precisely follow paths as directed by a computer program.

datasheet—A manufacturer-created description of an electronic component or assembly's functions, tolerances, and architecture.

DC motor—A commonplace motor that rotates its hub when voltage is applied to its terminals.

digital—A type of data that consists exclusively of yes-or-no instructions, versus analog data, which consists of varying voltage levels.

diode—An electronic component that typically allows voltage only in one direction.

encoder—A device that can detect how far a motor's hub has turned, and returns this value to a microcontroller.

flex sensor—A sensor built into soft plastic. It's essentially a variable resistor, with the resistance changing based on how far the plastic is bent.

Fritzing—Free electronics visualization software useful for designing circuits online. Look for a wiring diagram in this book and you'll see an example of Fritzing.

ground—The return path of an electric circuit. On a battery, the ground is marked with a - (minus sign). Ground is often abbreviated GND in electronic parlance.

ground bus—The strip of conductor on a breadboard, usually marked black or blue and designated as the ground.

hackerspace—*See* maker spaces.

heat-shrink tubing—Non-conductive rubber tubing used to cover wire joins. As heat is applied, the tubing shrinks down to cover up the exposed wire.

infrared (IR) light—A bandwidth of light outside of the visible range for humans, IR light is often modulated to send small amounts of data—for instance, the "off" signal for a TV.

integrated circuits (ICs)—A series of circuits miniaturized, then embedded in a plastic housing.

Integrated Development Environment (IDE)—Software that provides technical services to programmers to assist them in creating code.

interrupt pin—An Arduino pin that can interrupt a loop. If you wanted a button-push to stop a loop, you would need to wire up the button to an interrupt pin.

IR receiver—Sensor that detects infrared light pulsed at the correct frequency, 38 Mhz.

jumper—A generic term for wires or conductors used in electronics projects.

Kerf-bending—A laser-cutting trick that enables you to bend thin sheets of wood by making a series of cuts in the material.

keylock switch—A switch that turns with a key, allowing you to restrict who can activate your project.

knock sensor—A sensor that detects when it has been struck and sends voltage to the next part of the circuit.

laser cutter—Also known as a laser etcher, a laser cutter burns through thin materials such as cardboard, MDF, and particle board.

laser diode—The electronic module that emits a laser beam when voltage is applied to its terminals.

lathe—A device for shaping wood that works by rotating the material at high speeds while an operator applies a tool.

lead—A wire or terminal on a component to which a wire is attached.

LED—Short for Light Emitting Diode, the LED is the light bulb of the electronics world.

LED driver—An integrated circuit able to control multiple LEDs without maxing out the Arduino's pins.

library—Supporting code referenced by an Arduino sketch, allowing you to keep the main sketch relatively simple.

light sensor—A sensor that detects light. Some of these operate as a variable resistor, where the level or light dictates resistance, whereas others are digital and send numeric data to the microcontroller.

maker spaces—Communal workshops where tools and expertise can be shared, classes taught, and projects built.

MDF—Medium density fiberboard, an artificial wood that lends itself to maker projects.

mesh network—A network consisting of multiple nodes, each able to see every other node.

microcontroller—A miniature computer, able to take input from sensors and activate motors and lights.

motor control chip—An integrated circuit optimized to control motors, expanding on the Arduino's capabilities.

multimeter—A combination voltmeter and ohmmeter with additional functionality, designed to be the electrical engineer's primary measurement tool.

multitool—A folding tool, often in the form of pliers, with additional tools such as drivers, blades, scissors, and so on.

open-source hardware and software—Electronics projects where the code and electronic designs are shared freely, and anyone is free to modify or recreate it.

passive infrared (PIR) sensor—An infrared sensor that detects movements via subtle changes in temperature.

peristaltic pump—A pump that works by massaging a tube, preserving the substance pumped from contamination.

piezo buzzer—A component that buzzes when voltage is applied to its terminals.

pin—The power and data connectors of an Arduino.

plasma cutter—A CNC machine that cuts metal according to a design on a computer.

PVC—Polyvinyl chloride, also known as PVC, is the project-friendly plastic pipe most commonly used as plumbing pipes.

potentiometer—Usually referred to as pots, potentiometers are variable resistors adjusted by turning a knob.

power bus—The conductor strip on a breadboard designated to supply voltage to the board.

PowerSwitch tail—A convenient and safe tool for triggering wall-current with signals from a microcontroller.

pressurized reservoir—A way of pumping water by pressurizing a reservoir of liquid, forcing it out of an exit tube.

printed circuit board (PCB)—Composite boards coated in a conductive material, enabling you to etch circuits onto the board and thereby create electronic assemblies.

Real-Time Clock (RTC) module—A timekeeping chip with a battery backup, designed to maintain the correct time for several months.

relay—A microcontroller-triggered, electromechanical switch able to control high-voltage circuits.

resistor—An electronic component designed to limit the flow of electricity to protect fragile components and control the flow of voltage in the circuit.

RGB LED—A light-emitting module consisting of three elements, one each of red, blue, and green. By lighting one or more of these elements, a large variety of colors can be created.

rotary tool—A small power tool with multiple types of attachments ranging from saws to sanders to polishers. You've probably heard of the category leader, Dremel.

schematic—The drawn representation of a circuit, with symbols representing the various components.

sensor—An electronic device that sends data or voltage to a microcontroller about the environment around it.

serial communication—A method of data whereby data is sent along a single wire, with each bit sent sequentially.

serial monitor—The window in the Arduino IDE where serial traffic can be monitored. This can be a great tool for debugging programs.

servo—A motor equipped with a gearbox and encoder, enabling precision control of how far the motor's shaft turns.

seven-segment display—An LED display of a letter or number, formed out of seven smaller LED segments.

shield—An add-on circuit board for the Arduino. It stacks right on top, sharing the Arduino's pins while adding additional capabilities.

sketch—Arduino parlance for the program that controls the Arduino's pins.

solenoid—A motor, only instead of the shaft rotating, it moves back and forth. This is often used for valves.

standoffs—Metal or plastic inserts that create space or support between a PCB and another surface.

stepper motor—A motor designed to rotate in increments, called steps. It usually has four or more wires.

Sugru—Moldable, quick-setting adhesive putty with myriad uses.

table saw—A saw in the form of a work table with a saw blade sticking out of the surface.

temperature and humidity sensor—A digital sensor that measures temperature and humidity and returns a numeric reading to the microcontroller.

terminal strips—The rows of connectors in breadboards, running perpendicular to the power and ground bus.

tilt sensor—A sensor with a conductive ball rolling inside, so it knows when the sensor has been tilted to one side.

transistor—A miniature electronic switch controlled with electrical signals.

ultrasonic sensor—A sensor that detects obstructions and measures distances by transmitting a beam of inaudible sound and then listening for an echo.

voltage regulator—A component that helps measure the right amount of voltage in an electronics project.

XBee—A wireless module using the popular Zigbee protocol, which is often used for home automation.

Index

Symbols

A

B

G

H

I

J

N

O

P-Q

R

S

T

U

V

W

X-Y

Z